NOT A CHIMP

Jeremy Taylor is a science documentary film producer and director who has spent a large proportion of his professional career in the Science Department of BBC Television. Between 1981 and 1991 he spent much of his time on the BBC's science television flagship series *Horizon*, for which he made over a dozen one-hour films. Since 1991, as a freelance television producer, he has been a regular contributor to science on Channel 4 in the UK, and, through them, the Discovery Channel and National Geographic Television. He has won several top industry awards for his film-making including a Special Merit Citation at the 1985 Chicago International Film Festival; Best Science Film at the 1987 Royal Society International Scitech Film Festival; and Silver Awards for Educational Merit and Medicine In The Media at the 1999 British Medical Association Film Competition. Many of his films have been on biological subjects and have stressed the role of evolution. They include *Playing with Madness, Mindreaders,* and *Nice Guys Finish First* and *The Blind Watch-Maker* with Richard Dawkins. This is his first book.

http://notachimp.blogspot.com

NOT A CHIMP

The hunt to find the genes that make us human

JEREMY TAYLOR

OXFORD
UNIVERSITY PRESS

OXFORD

UNIVERSITY PRESS

Great Clarendon Street, Oxford OX2 6DP

Oxford University Press is a department of the University of Oxford.
It furthers the University's objective of excellence in research, scholarship,
and education by publishing worldwide in

Oxford New York

Auckland Cape Town Dar es Salaam Hong Kong Karachi
Kuala Lumpur Madrid Melbourne Mexico City Nairobi
New Delhi Shanghai Taipei Toronto

With offices in

Argentina Austria Brazil Chile Czech Republic France Greece
Guatemala Hungary Italy Japan Poland Portugal Singapore
South Korea Switzerland Thailand Turkey Ukraine Vietnam

Oxford is a registered trade mark of Oxford University Press
in the UK and in certain other countries

Published in the United States
by Oxford University Press Inc., New York

First published 2009
First published in paperback 2010

British Library Cataloguing in Publication Data

Data available

Library of Congress Cataloging in Publication Data

Data available

Taylor, Jeremy, 1946-
Not a chimp : the hunt to find the genes that make us human / Jeremy Taylor.
p. cm.
ISBN 978-0-19-922778-5 (Hbk.)
ISBN 978-0-19-922779-2 (Pbk.)
1. Human genetics–Popular works. 2. Human evolution–Popular works. 3. Animal
behavior–Popular works. 4. DNA–Popular works. I. Title.
QH431.T296 2009

599.93'5–dc22   2009013846

Typeset by SPI Publisher Services, Pondicherry, India
Printed in Great Britain
on acid-free paper by
Biddles Ltd., King's Lynn, Norfolk

ISBN 978-0-19-922778-5 (Hbk.)
ISBN 978-0-19-922779-2 (Pbk.)

1 3 5 7 9 10 8 6 4 2

For Wilma,
who did not live quite long enough to see this book published, and
Barbara, without whom it could not have been written

PREFACE

In many ways, this book is born out of frustration for a professional career in popular science television where ideas about comparative primate cognition, and the similarities and differences between us and our primate relatives, have continually circled me but constantly evaded my grasp in terms of the opportunity to transform them into science documentary. On the plus side, keeping a watching brief for over a quarter of a century on subjects like comparative animal cognition and evolution allows you to watch a great deal of water flow under the bridge. Fashions come and fashions go—specifically, perspectives on the similarity—or otherwise—of human and ape minds.

I remember the first *Horizon* science documentary about the chimpanzee Washoe, the great ape communicator, using American Sign Language to breach the species barrier. And, later, Kanzi the bonobo, jabbing his lexicon. These were the apes, as Sue Savage-Rumbaugh put it, that were 'on the brink of the human mind'.

I remember when the pre-print of *Machiavellian Intelligence*, by Andrew Whiten and Dick Byrne, plopped onto the doormat of the BBC *Antenna* science series office in 1988. Suddenly primatology had become a great deal more exciting. Could primates, and particularly higher primates like chimpanzees, really be as full of guile, as dastardly, as cunning, and as manipulative as the eponymous Florentine politician? Could they really reach deep into the minds of other individuals to see what they believed and what they wanted, and turn that information into deception?

I remember discussing primate cognition with a young Danny Povinelli, as we sat finger-feeding ourselves shrimp gumbo and new potatoes out of plastic Tupperware containers in a Lafayette restaurant surrounded by an alligator-infested moat, before returning to his kingdom—the New Iberia Research Centre—where the University of Louisiana had lured him back to his native deep South by turning a chimpanzee breeding centre for medical laboratory fodder into a primate cognition laboratory with one of the largest groups of captive chimpanzees in the country. He looked like a kid who had just been thrown the keys to the tuck shop.

In those days Povinelli shared the *zeitgeist*—spread by Whiten's and Byrne's work, and started by Nick Humphrey and Alison Jolly before them—that, since the most exacting and potentially treacherous environment faced by chimpanzees and other primates was not physical, but the social environment of their peers, they had evolved a form of social cognition very much like our own, in order to deal with it. This was further elaborated into a full-blown 'social brain' hypothesis by Robin Dunbar, who related brain neocortex size to social group size throughout the primates and up to man. Povinelli's early work reflects this optimism for the mental life of apes, but both ape-language and ape-cognition research were subjected to a cold douche of searching criticism during the 1990s, and misgivings set in regarding the effectiveness of the experiments that had been constructed to gauge ape cognition. Now the worm has turned again, with a number of research groups emerging with bolder and bolder claims for the Machiavellian machinations of primate minds, only to be powerfully countered by the curmudgeonly scepticism, chiefly by Povinelli, that these researchers are merely projecting their mental life onto that of their subjects; that, rather in the frustrating manner of Zeno's arrow that could never quite reach its target because it continually halved its distance to it, no experiment constructed thus far can actually get inside the mind of a chimp and show us exactly what it does and doesn't know, or how much, about the minds of others or the way the physical world works. One influential part of the world of comparative animal cognition talks of a continuum between ape and human minds and shrinks the cognitive distance between us and chimps to almost negligible proportions, while another returns us to the unfashionable idea that human cognition is unique, among the primates, after all.

When I began writing this book the working title was 'The 1.6% that makes us human'. My aim had always been to scrutinize the impression put about in the popular science media that humans and chimps differ by a mere 1.6% of our genetic code—or even less—and that it therefore makes complete sense that this minuscule genetic difference translates into equally small differences in cognition and behaviour between apes and man. However, contemporary genome science and technology, over the last few years, have dramatically advanced the power and resolution with which scientists can investigate genomes, eclipsing the earlier days of genome investigation that gave rise to the '1.6% mantra'.

As with comparative cognitive studies, conclusions on chimp–human similarity and difference in genome research depend crucially on perspective. To look at the complete set of human chromosomes, side by side with chimpanzee chromosomes, at the level of resolution of a powerful light microscope, for instance, is to be overwhelmed by the similarity between them. Overwhelmed with a sense of how close our kinship is with the other great apes. True, our chromosome 2 is a combination of two chimp chromosomes—giving humans a complement of 23 chromosome pairs to 24 in chimps, gorillas, and orang-utans—but even here, you can see exactly where the two chimp chromosomes have fused to produce one. The banding patterns you visualize by staining the chromosomes match up with astonishing similarity—and that banding similarity extends to many of the other chromosomes in the two genomes. However, look at a recent map of the chromosomes of chimps and humans, aligned side by side, produced by researchers who have mapped all inversions—end-on-end flips of large chunks of DNA—and the chromosomes are all but blotted out by a blizzard of red lines denoting inverted sequence. Now you become overwhelmed by how much structural change has occurred between the two genomes in just 6 million years. True, not all inversions result in changes in the working of genes—but many do—and inversions might even have been responsible for the initial divergence of chimp ancestor from human ancestor.

The extent to which you estimate the difference between chimp and human genomes depends entirely on where you look and how deeply. Modern genomics technology has led us deep into the mine that is the genome and has uncovered an extraordinary range of genetic mechanisms, many of which have one thing in common. They operate to promote variability—they amplify differences between individuals in one species. We now know, for instance, that each human is less genetically identical to anyone else than we thought only three years ago. When we compare human genomes to chimpanzee genomes these mechanisms magnify genetic distance still further. I have tried, in this book, to follow in the footsteps of these genome scientists as they dig deeper and deeper into the 'Aladdin's Cave' of the genome. At times the going gets difficult. Scientists, like any explorers, are prone to taking wrong turnings, getting trapped in thickets, and covering hard ground, before breaking through into new insights. I hope that those of you who recoil from genetics with all the visceral horror with which many regard the sport of pot-holing will steel yourselves and follow me as far as I have dared to go into Aladdin's Cave. For only then will you see the riches within and begin to appreciate, as I have, just how limited popular accounts of human–chimpanzee genetic difference really are.

Let me try and persuade you that this is a journey, if a little arduous at times, that is well worth taking. There are a number of scientists around the world who have the breadth and the vision to have begun the task of rolling genetics, comparative animal cognition, and neuroscience into a comprehensive new approach to the study of human nature and this is part, at least, of their story. They strive to describe the nature of humans in terms of the extent to which we are genuinely different to chimpanzees and the other great apes. Somehow, over 6 million years, we humans evolved from something that probably resembled a chimpanzee (though we cannot yet be entirely sure) and the answer to our evolution has to lie in a growing number of structural changes in our genome, versus that of the chimpanzee, that have led to the evolution of a large number of genes that have, effectively, re-designed our brains and led to our advanced and peculiar human cognition. If you don't believe me, hand this book to your nearest friendly chimpanzee and see what he makes of it!

ACKNOWLEDGEMENTS

A great number of busy, working scientists have taken time to answer my many questions, and to provide information and copies of scientific papers at my request. I am very grateful to them all. However, I am deeply indebted to a number of scientists who have been particularly generous with their time, information, help, advice, and patience. These are: Simon Fisher, Faraneh Vargha-Khadem, Marcus Pembrey, Jane Hurst, Geoff Woods, Christopher Walsh, Ajit Varki, James Sikela, Todd Preuss, Daniel Povinelli, Josep Call and colleagues at the Max Planck Institute for Evolutionary Anthropology, Nathan Emery, Nicola Clayton, and Alex Kacelnik. And especially Richard Wrangham, who suggested to me that the idea of human self-domestication was ripe for re-exploration, and thus kick-started chapter 11. This book is all the richer for their assistance while, needless to say, any remaining errors of fact or omission remain entirely my own.

Finally, I want to extend special thanks to my editor at Oxford University Press, Latha Menon. Having commissioned the book in the first place, she guided me through the development of my argument and the detailed editing that, hopefully, has brought it to a readable conclusion, with warm and energetic encouragement and the underlying steel of a gimlet eye for grammar and clarity.

CONTENTS

ILLUSTRATION LIST

ABBREVIATIONS

ADHD	Attention Deficit Hyperactivity Disorder
fMRI	functional MRI
HAR	Highly Accelerated Region of genetic code, showing significant mutation
KE	Family in west London suffering from SLI
MCPH	Genetic locus associated with MicroCePHaly
MHC	Major Histocompatibility Complex
MPFC	Medial Prefrontal Cortex
MRI	Magnetic Resonance Interferometry (a brain-imaging technique)
PET	Positron Emission Tomography (a brain-imaging technique)
Siglecs	Sialic acid-recognizing lectins
SLI	Specific Language Impairment
snip	Single nucleotide polymorphism

CHAPTER 1

From Distant Cousins to Close Family

An American tourist becomes lost in the wilds of Western Ireland. Spotting a farmer working in a nearby field, he edges his car toward the kerb, gets out, and walks over to him. 'Excuse me, sir', he says, 'Could you possibly tell me how to get to Kilkenny?' The farmer ruminatively scratches his stubbly chin, before answering: 'Well now, if I were going to Kilkenny, I wouldn't start from here!'

One of mankind's greatest intellectual journeys has been the quest for our evolutionary human origins. Increasingly, we search in our genes and the genes of other species for vital clues to where we come from and what makes us human. And, like the hapless traveller in the Irish landscape, we have not chosen the ideal place from which to start, for we have chosen the chimpanzee. We have had no choice in the matter because the modern coordinates on the map of our evolutionary history are marked in DNA and the chimpanzee is our nearest living DNA relative. It would have been preferable to set off gradually down the road into our evolutionary past, following our DNA way-points one step at a time, checking our bearings. But this is not possible. After the Neanderthals, a mere 35,000 years ago, the DNA simply peters out. Like the sign-posts in the maze of Irish country roads that so befuddled our tourist, the DNA has rotted away.

How helpful it would have been if there had still existed a knot of *Homo erectus* in a remote enclave in the Caucasus, or a carefully nurtured tribe of *Homo heidelbergensis* in an ecological preserve in southern Spain, or if the 'Frodos' of *Homo floresiensis* were still valiantly slaying Komodo dragons in Indonesia. But they have all gone, and their DNA and behaviour has gone with them. All we have is a jigsaw of hominin fossils, which, after well over a century of field work, is still so surprisingly small and fragmented that one wag, rather stretching his point, once exclaimed that the whole lot could be contained within a decent-sized wheelbarrow. The jigsaw comes without the big picture and with most of the pieces missing.

Over the past fifteen years we have seen an explosive growth in the power of technologies that allow us to sequence vast tracts of DNA in our genome, culminating in the first rough draft of the human genome in 2000, and a more detailed, polished version, in 2003. Since 2005 we have also had a good quality read of the chimpanzee's 'Book of Life'. This has allowed us to leapfrog back in time, over the heads of our more immediate hominid ancestors, to compare human genes with chimpanzee genes. It means that we have begun in earnest the task of defining what makes us human by virtue of the genetic differences between us and a species from whom we became separated approximately 6 million years ago. And, since chimpanzees have also been evolving, this means that 12 million years of evolutionary time separate us—6 million for each branch of the tree since the split from the common ancestor.

Although this massive and audacious scientific exercise makes sense to the myriad genome scientists it employs, it has also given rise to a number of fallacies, including the assumption that we are descended from the chimpanzee when we are not, and the idea that our humanness may be traceable to a handful of mutated genes among those relatively few that appear to differ between us and apes. It has also led to a very blinkered view of human evolution due to exclusive comparison between humans and chimpanzees. We also find ourselves in the unfortunate position of adopting chimpanzees as the most important reference point for charting our evolution when they are in grave danger of falling off the map altogether—of becoming extinct.

This obsession with ape–human comparisons began when pre-Victorian explorers first brought back to Europe their disconcerting accounts and examples of the bewildering variety of monkeys, apes, and sub-Saharan Africans retrieved from 'The Dark Continent'—West Africa. In the days well before palaeontology had begun making major contributions to studies of human origins, this rich broth of hominoidae, with feral children and the mentally sub-normal thrown in for good measure, was *the* food for thought about the relationship between man and the animals. How best to order Nature, and chart the origin of faculties like language and the propensity for civilization and culture that were thought to define human beings? Here, in the seventeenth century, are the roots of anthropology and primatology, and the acknowledgment that asking questions of the great apes would lead to an understanding of the nature of 'humanness'.

According to Thomas Huxley, the first coherent descriptions of the great apes began coming out of Africa during the 17th century. Particularly colourful are the accounts of Andrew Battell, published around 1625, of a monstrous creature encountered during his extensive exploration of Angola:

> The Pongo is in all proportion like a man; but that he is more like a giant in stature; he is very tall, hath a man's face, hollow-eyed, with long haire upon his browes. His face and eares are without haire; and his hands also. His body is full of haire, but not very thicke; and it is of a dunnish colour.

Battell's beast, clearly a chimpanzee, couldn't speak, slept in trees, built shelters from the rain, ate fruits, and had a nasty habit of attacking and killing Africans as they worked in the forest.

Were these apes primitive humans, or another animal group? Were they capable of language? Soon, live specimens were making their way back to the UK. In 1661, a large primate from Guinea certainly had Samuel Pepys fooled:

> I do believe it already understands much English: and I am of
> the mind that it might be taught to speak or make signs.

In 1698 a young chimpanzee that had died from infection soon after its voyage from Africa was dissected by the anatomist Edward Tyson. Tyson, a man very much before his time, and a scientific pioneer of comparative ape–human studies, published an exhaustive account of the similarities and differences between the anatomies of ape and human, noting 47 anatomical similarities. He was particularly surprised by the apparent similarity between the chimpanzee and human brain.

In the 18th century, natural philosophers resurrected the Aristotelian idea of a *scala naturae*—a linear Great Chain of Being running from the lowest forms to man—from religious dogma and applied it across the mineral, plant, and animal kingdoms up to monkeys, apes, feral children, black Africans, and modern Europeans. Man was the God of this Chain of Being, lower only than the Deity Himself. This *scala* was the forerunner of a theory of evolution. Foremost among these thinkers was the extraordinary Scottish peer Lord Monboddo who declared that man had progressed from an animal, through an orang-utan (he could have meant any ape—the terms were interchangeable in those days), and he was convinced that the only reason orang-utans had not acquired language was because they had not yet advanced enough culturally.

Carolus Linnaeus, the great Swedish taxonomist, produced the first grand classification of all plants and animals known on Earth.

He had enormous trouble incorporating this flood of humans, apes, and in-betweens from the New World and even made room for mythical animals in his classification. He rejected a simple, linear Chain of Being by broadening its vertical axis to create Groups which split into Genera and Species: for instance, the Primates, in which group he placed humans, monkeys and apes, lemurs, and bats. He first grouped chimpanzees and orang-utans into the species *Homo troglodytes*, and dubbed us *Homo sapiens*. However his schema was still, in essence, a grand *scala naturae*. All species were fixed, immutable, graded vertically according to their external characteristics, knew their place, and were made by God.

Darwin should have obliterated the idea of the Great Chain of Being, the *scala naturae*, with his publication of *The Origin Of Species* in 1859, which fatally challenged ideas of the fixity of Nature—and he was quite clear that humans had evolved with chimps from a common ancestor. Both Darwin and Thomas Huxley sought to minimize the cognitive distance between man and the apes, the better to resist clerical criticism from churchmen like 'Soapy Sam' Wilberforce. Language was the stumbling block for Darwin, because neither he nor Huxley could offer a persuasive account of how language could have evolved in the tiny, gradualistic steps he claimed typified evolution by natural selection. This opened the door for his main critic, Max Muller, of the University of Oxford, who threw down a gauntlet that we are still facing today, when he exclaimed:

> The one great barrier between the brute and man is language.
> No process of natural selection will ever distill significant words
> out of the notes of birds or the cries of beasts.

Today we are still asking the same two linked questions that so obsessed those earlier natural philosophers: How did we evolve from the apes? How did we become human? Since the days of these pre-Victorian pioneers we have added over a century of palaeontological research, decades' worth of field primatology, and powerful modern laboratory techniques of comparative genetics, neuroscience, and cognitive psychological testing. In Victorian times free-thinking natural philosophy crossed swords with religion. Ironically, religion, in the form of creationism and intelligent design, still mounts a challenge to orthodox evolution for those credulous enough to find its arguments convincing. But today we have to add another factor that has muddied the water and muddled the thinking on our evolutionary relationship to the apes. It is the very imminent threat to the existence of chimpanzees in the wild. This has led to the drastic need to safeguard ape species and conserve their habitat, and conservationists have searched for the most potent argument to galvanize us into action. All this has led some scientists, intentionally and unintentionally, to exaggerate the narrowness of the gap between chimpanzees and ourselves because it is argued that if the genetics and neurobiology of chimps is virtually indistinguishable from our own then so must be their behaviour. And if they are one of us, then we are morally bound to their welfare. This is one version of the so-called argument from analogy. It is a view, I will argue, that is as wrong as it is misguided. It plays into the hands of our natural propensity to anthropomorphize our pets and other animals, and even our inanimate possessions, and it has allowed us to distort what the science is trying to tell us. I want to set the record straight and restore chimpanzees to arm's length. I also want to show how science over the past few years really is beginning to point the way to how we became human and how much, and in what way, we are different to chimpanzees. I will do it using three weapons from the sciences that compare chimpanzees to humans: genomics, neuroscience, and cognitive psychology. First, however, let us examine our relationship to the chimpanzee a little further.

At the beginning of March 2005, St. James and LaDonna Davis travelled to the Animal Haven Ranch in Caliente, California to celebrate the 39th birthday of Moe, a male chimpanzee they had previously owned and raised at home. They had arrived with lots of new toys for Moe and a 'beautiful sheet cake with raspberry filling'. LaDonna first spotted, out of the corner of her eye, that one of the neighbouring chimps was free, just as she was about to cut the cake. CNN reported what happened next:

> The couple had brought Moe a cake and were standing outside his cage when Buddy and Ollie, two of four chimpanzees in the adjoining cage, attacked St. James Davis. He sustained severe facial injuries and his testicles and a foot were also severed. Buddy, a 16 year old male chimp, initiated the attack and after he was shot, Ollie, a 13 year old male, grabbed the gravely injured man and dragged him down the road.

A local TV station, NBC4, reported LaDonna Davis saying that the attack swiftly overwhelmed her husband:

> When we made eye contact, the charge was on. There was no stopping anything, and the big chimp came around from behind me and pushed me into my husband. The male came around from behind and chomped off my thumb ... my husband must have realized we were in deep trouble because he pushed me backward. At that time, they both went for him. One was at his head, one was at his foot. But all the time *he was trying to reason with them*. They virtually were—I don't know how you say it—eating him alive.

They further reported that, in addition to his testicles and one foot, St. James Davis lost all the fingers from both hands, an eye, part of his nose, cheek, lips, and part of his buttocks in the ferocious attack. He was removed to the Loma Linda Medical Center where he was placed into an induced coma while he fought for his life.

A couple of months later, thanks to the *Washington Post*, details of the Davises' relationship with Moe emerged. St. James had retrieved the baby chimp from Tanzania in 1967, before the two got married. He was a former Nascar driver and would carry the infant around his auto body-shop in West Covina, in a sling. Moe was their illegitimate child and best man at their wedding. He slept in the Davises' bed until he got too big. He learned to use the toilet. He used to love watching Westerns on TV. Cancer had left LaDonna infertile. Moe was the son they could never have.

Unfortunately, a shadow was eventually cast over the Davises' domestic bliss. Moe escaped from his enclosure and rampaged through the neighbourhood, biting a police officer's hand while under restraint. He later bit a woman's finger off when, against their warning, she had poked it, with her red-painted fingernails, through the bars of Moe's cage. The Davises claimed he had mistaken it for liquorice. City officials stepped in and demanded Moe's instant removal to the animal sanctuary. They were devastated.

By May 7th *Racing West* was able to publish a follow-up about the Nascar veteran, who, by then, had been brought out of coma:

> He made an effort to speak but it was nearly impossible because his lips were missing. Doctors gave him an electronic voice box, to hold against his throat, and his first ever words were 'How is Moe doing?'

This story left a lasting impression on me because nothing I have ever heard so graphically demonstrates the ambivalent world of chimpanzee–human relationships: huge emotional attachment of human to chimp; bizarre levels of anthropomorphizing; an animal species capable of thrilling us with its human-like behaviour on the one hand and horrifying us with its brutal aggression on the other. And a man appealing, verbally, to a chimpanzee's better nature while it was in the act of emasculating him.

Why is it that so many of us find it so important to identify ourselves with the chimpanzee—even when nature bites back? Why is it that we are encouraged to identify with them in order to support their ecological salvation? Why is it that human uniqueness has become such a shameful idea? Is it that we, as laymen, simply cannot prevent our innate ability to anthropomorphize chimps from cloaking them in our cognitive clothes? Is it that we, as biologists, are so petrified of creationists that we not only stress the biological continuum between us and chimpanzees, but seem to find it obligatory to collapse that distance into close proximity? Does it involve an exaggeration of the sagacity and wisdom of chimps, a loss of confidence in where the power of human intellect has got us, or both?

Before genome science allowed us to open up a second front in our quest for human origins, we relied upon palaeontology. Back in the 1950s, the world-famous palaeontologist Louis Leakey acknowledged two key problems with this approach, which are that neither DNA nor behaviour fossilizes, forcing us to deduce the behaviour of our ancestors indirectly from what clues we can garner from their fossilized skeletal remains. This is why he decided that man's origins needed collateral investigation through the field study of the three major great apes. He hoped it would provide crucial insights into how our evolutionary ancestors behaved. Certainly, much of what we now know about the behaviour of great apes in the wild was spawned by the field work of Jane Goodall on chimpanzees, Dian Fossey on gorillas, and Birute Galdikas on orang-utans, Leakey's 'angels'.

I would not dispute for a moment the value of much of this field research. It has told us a great deal we did not know about primate behaviour and social organization. What is more in question is how much it has told us about the evolution of human behaviour. It is quite clear what Jane Goodall's Institute thinks:

> Various mental traits once regarded as unique to humans have been convincingly demonstrated in chimpanzees; reasoned thought, abstraction, generalization, symbolic representation, and concept of self. Non-verbal communications include hugs, kisses, pats on the back, and tickling. Many of their emotions, such as joy and sadness, fear and despair, are similar to or the same as our own. Once we admit that we are not the only beings with personality, reasoned thought, and above all, the ability to feel and express emotions such as joy, despair and empathy,

then we develop a new respect for chimpanzees. *The line between human and other non-human beings, once thought so sharp, becomes blurred.* (emphasis added)

That line has sometimes seemed close to disappearing altogether. We have Jane Goodall to thank for the first recorded instance of chimpanzee tool use. Her observations, in 1960, that one of her favourite male chimpanzees at Gombe, David Greybeard, was using grass stems to fish for termites in a mound, licking off the insects that clung to the probe, led to Louis Leakey's famous overreaction: 'Now we must re-define Tool, re-define Man, or accept chimpanzees as humans.' Goodall further observed her chimpanzees breaking off twigs from trees and stripping them of their leaves to fashion a more suitable fishing tool. Since then, chimpanzees across Africa have been observed using twigs to dip for honey and probe for bees, termites, and algae; have been seen to use leaves to sponge up liquid; crack nuts against hard surfaces, or by using crude stone hammers and anvils; pound nuts as if using a crude mortar and pestle; and use simple hooks, picked out of vegetation, to forage for insects. They have recently been observed using sharp-tipped branches as crude spears, using sticks as clubs, and throwing rocks at retreating prey. There is little doubt as to where Goodall now stands on the relationship of humans to chimps. Her famous movie is titled *Chimps: So Like Us*. For Goodall, it seems, chimps are 'virtually human beings'.

Scientific talk about chimp–human proximity boils over readily into popular human culture. James Mollison's beautiful portraits of chimpanzees and the other great apes invite us to 'Look Into My Eyes' to see ourselves reflected back. In 2006, the BBC's flagship science television series *Horizon* aired a programme, fronted by the comedian Danny Wallace, called 'Chimps Are People Too'. 'Danny is on a mission', ran the blurb, 'to convince the world that chimps are people too. He believes the time has come to make our hairy relatives part of the family.' In early 2008, the UK's Channel 5 showed a documentary film on Tetsuro Matzuzawa's chimpanzee research in Japan, which had already drawn a blizzard of 'Chimp beats college students at math' newspaper headlines. The film documented Matzuzawa's star chimpanzee, Ayuma, beating humans at a computer game where all the numerals 1 to 9 are flashed up in boxes in random pattern on the screen, for a minute length of time, before being replaced by the empty boxes. The subject has to touch the screen to re-conjure the numerals in order. Chimps are very good at this—better than humans—and typically remember up to 80% of the numerals even when the original pattern is displayed for less than half a second. 'The human race may be about to lose their reputation as the cleverest species on the planet!' sang the commentary. At last, the sheer arrogance of humans, in their assumption of superiority to other apes, had been exposed! Such was the implication. Yet this was a test of eidetic—or photographic—memory, not mathematical skills. Tiny children have this skill before it becomes engulfed by language and a genuine symbolic understanding of numerals. Idiot savants have excellent eidetic skills alongside profound cognitive deficits. In what real sense, then, has the chimpanzee leapfrogged us?

Evidence from the wild on chimpanzee tool use and the way they play tricks on each other—appearing to intentionally deceive each other—has been presented as clear evidence that the barrier of human uniqueness has been fatally breached. Yet, as we shall see, cognitive scientists today are hopelessly divided over how much a chimpanzee knows about the nature of physical forces in the world when it shapes its crude tools and how much it knows about the mental life of other individuals when it interacts with them.

Just as field primatology has been interpreted so as to close the gap between humans and chimpanzees, so too has evidence from protein chemistry and genome science. Prior to 1960, anthropology had relied on traditional methods of classification to design taxonomic trees representing the relationships between the various genera, families, and orders of the primates. This traditional method uses the concept of grades of evolutionary advancement based on a distinction between primitive and more advanced features to group primates together. Brain size would be one such feature. Thus, all the great apes were classified together in the *Pongidae*, and humans were seen as a special and more advanced case. To Morris Goodman, then at the Detroit Institute of Cancer Research, this seemed suspicious and he set about re-drawing the primate family tree.

Conventional wisdom at the time had gibbons first splitting off from a branch to the common ancestor of all the other great apes and humans. The next oldest split separated the human line from chimps, orang-utans, and gorillas, while subsequent events first split off the orang-utan, and then separated the chimpanzee and the gorilla. Goodman queried this taxonomic remoteness of humans to the rest of the apes, and the close relationship drawn between gorillas and chimpanzees. These were primitive days for molecular biologists. It was less than a decade since the structure of DNA had been discovered by Crick and Watson, there was no technology for sequencing DNA, nor sequencing amino-acids in the proteins that DNA codes for. He discovered that human serum proteins reacted far more strongly with those of gorilla and chimpanzee than with orang-utan and gibbon—suggesting that they were much more closely related. He presented his results at a symposium called 'The Relatives of Man' in 1962, arguing that his fledgling science of molecular taxonomy would logically place chimps and gorillas alongside Man in the *Hominidae*. The idea was brusquely thrown out by the great evolutionary biologists George Gaylord Simpson and Ernst Mayr. To Goodman, these traditional biologists were perpetuating an anthropocentric view of nature, hearkening back to the dark old days of the pre-Darwinian idea of a *scala naturae* with Man at the top of the ladder. He rushed back to Detroit to develop his new science. Early computer programmes now existed for comparing amino-acid and limited amounts of DNA nucleotide sequence data and he developed these into a way of grouping animal species by how close they were in terms of protein sequence. By the early 1980s he had compared the sequences of six types of globin protein, and four further proteins; two fibrinopeptides, cytochrome C, and an enzyme, carbonic anhydrase. *Homo* and *Pan* (the chimpanzee) showed 99.6% identity in amino-acid sequence versus 99.3% identity between *Gorilla* and either *Homo* or *Pan*. Further, the genetic distance between chimpanzees and humans was less than that between chimpanzees and gorillas. The conclusion was that chimps and humans belonged together in the same genus. By then, several research groups had calculated, based on the rate of DNA nucleotide substitution over time, a molecular clock of evolution which put the divergence of chimpanzees from the human line at 5.9 million years, whereas previous estimates, based on fossil evidence, had put it at 20 million. Our distant cousins were rapidly becoming part of our nuclear family.

In his 1989 book *The Rise And Fall Of The Third Chimpanzee* Jared Diamond picked up this molecular data and ran with it. He invited us to consider an arresting thought experiment. Create a new enclosure in your local zoo, he suggested, and fill it with a number of human beings who have been stripped naked and reduced to visceral grunts. A visitor from outer space, armed with reasonable powers of observation, 'would have no hesitation in seeing us for what we are—chimps with little hair that walk upright—and would immediately classify us as a third species of chimpanzee, along with the pygmy chimp of Zaire and the common chimp of the rest of tropical Africa'. The chimp had the genetic right to be called *Homo troglodytes*, or, in another way of establishing a level genetic playing-field, we should be renamed *Pan sapiens sapiens*.

Goodman went on, in the late 1990s, to argue that, since humans appeared to share more than 98.3% of their non-coding DNA with chimps, and likely would be found to share up to 99.5% within genes actively coding for protein, we should see humans as little more than slightly remodelled apes. All apes should join humans in the family *Hominidae*, and chimps and the closely related pygmy chimp, or bonobo, should join humans in the genus *Homo*. In 2000, he joined forces with a number of prominent American scientists with interests in molecular evolution to lobby for a Human Genome Evolution Project that would allow scientists to directly compare the entire human genome (then only months away from publication) to that of chimpanzees, bonobos, and more distantly related higher primates. In their proposal they noted that the behavioural distance between humans and chimps was much smaller than originally thought and the genetic distance minuscule. They also noted that the New Zealand parliament was considering legislation, based on scientific advice and those interested in animal rights, that all hominids be given three basic legal rights: the right not to be deprived of life, the right not to be subjected to torture or cruelty, and the right not to be subjected to medical or scientific experimentation. Since, thanks to Goodman, there was a good argument for including chimpanzees in the genus *Homo*, that could mean extending human rights to chimps.

Between 2001 and 2005, four important scientific studies confirmed Goodman's earlier work on human–chimp DNA sequence similarity. In 2001, Feng-Chi Chen and Wen-Hsiung Li, from National Taiwan University and the University of Chicago, compared 53 regions of DNA that did not contain any known genes, between all the great apes. Earlier work, like Goodman's, they reminded us, had relied on cruder techniques and very localized areas of DNA and had put the difference between the chimpanzee and human genomes at about 1.6%. To increase confidence, many areas ought to be sampled, and genes, which might have been subject to natural selection that might locally skew the difference, should be avoided. They found that sequence divergence between chimp and human was only 1.24%. In 2002, Ingo Ebersberger and colleagues at the Max Planck Institute for Evolutionary Anthropology in Leipzig, compared 8,859 DNA sequences from all over the genome, totalling around 2 million pairs of DNA nucleotides, between chimpanzee and human. They also put the average sequence divergence at 1.24%. In 2004, Derek Wildman, Morris Goodman, and colleagues from Wayne State University took 90,000 DNA base-pairs of coding sequence (part of the small proportion of the genome that actually contains functional DNA in the form of genes for making proteins), which represented parts of 97 human genes, and compared them to their chimpanzee equivalents. They were 99.4% similar. Again, they stressed the behavioural similarity between the two species, which appeared logical when you considered their genetic proximity. Finally, in 2005, the Chimpanzee Sequencing and Analysis Consortium, a massive collaboration of 67 top genome scientists representing 24 laboratories world-wide, published their comparison of the initial complete sequence of the chimpanzee genome with its human counterpart. This meant comparing over 20,000 shared genes and 3 billion base-pairs of DNA. Over 96% of each genome was directly comparable and within those regions, on average, the two genomes were 98.8% similar. Nevertheless, they noted, even such high sequence similarity accommodated 35 million instances where a single DNA nucleotide had been substituted between the two species, 5 million sites where larger substitutions were present, and some loss or gain of genes by duplication or deletion, which brought the overall total sequence similarity down to 96%.

Naturally, *Science* magazine made the discovery their 'breakthrough of the year'. Elizabeth Culotta and Elizabeth Pennisi noted that now, armed with both sequenced genomes, scientists could examine, one by one, the 40 million evolutionary events that separated humans from chimps. 'Somewhere in this catalog of difference lies the genetic blueprint for the traits that make us human: sparse body hair, upright gait, the big and creative brain.' One of the senior Consortium scientists, Svante Paabo, said: 'I'm still sort of taken aback by how similar humans and chimps are in their DNA. I'm still amazed, when I see how special humans are and how we have taken over the planet, that we don't find stronger evidence for a huge difference in our genomes.' The Consortium saluted Darwin:

> More than a century ago Darwin and Huxley posited that humans share recent common ancestors with the African great apes. Modern molecular studies have spectacularly confirmed this prediction and have refined the relationships, showing that the common chimpanzee and pygmy chimpanzee are our closest living evolutionary relatives. Chimpanzees are thus especially suited to teach us about ourselves, both in terms of their similarities and differences with humans.

They concluded by ruing the fact that the very existence of chimpanzees was threatened just at the time they were helping us establish the genes that made us human. Their genetic proximity to us should be a clarion call for action to preserve them:

> More effective policies are urgently needed to protect them in the wild. We hope that elaborating how few differences separate our species will broaden recognition of our duty to these extraordinary primates that stand as our siblings in the family of life.

Accompanying the publication of the paper with Derek Wildman, Goodman was quoted in *National Geographic* as saying:

> The loss of the wild chimp and gorilla seems imminent. Moving chimps into the human genus might help us to realize our very great likeness, and therefore treasure more and treat humanely our closest relative.

From the work of the early pioneer of chimp–human proximity, Morris Goodman, up to the conclusions of the Genome Consortium in 2005, increasing evidence for the genetic similarity of humans to chimpanzees has been allowed to suggest that it is logically mirrored in terms of behavioural similarity, and these scientific conclusions have been conflated with preservation issues by arguing, in effect, that chimpanzees, and the other great apes, are a special case because they are virtually human beings. For some years this has been enshrined in the philosophy of the Great Ape Project, started in 1993 by the philosophers Peter Singer and Paula Cavalieri. The logo on their website, showing the tips of a chimpanzee finger and a human finger nearly meeting through the bars of a cage, is a crude pastiche of Michelangelo's 'Creation of Adam', from the ceiling of the Sistine Chapel, which portrays God imparting the spark of life to humanity. Is this simply used as a symbol of brotherhood, or, in a reversal of the geometry of the original painting, is the chimpanzee imparting the spark of humanity to the human? This is their declaration:

> We demand the extension of the community of equals to include all great apes: human beings, chimpanzees, bonobos, gorillas and orang-utans.
>
> The community of equals is the moral community within which we accept certain basic moral principles or rights as governing our relations with each other and enforceable at law. Among these principles or rights are the following: The right to life, the protection of individual liberty, and the prohibition of torture.
>
> The idea is founded upon undeniable scientific proof that non-human great apes share more than genetically similar DNA with their human counterparts. They enjoy a rich emotional and cultural existence in which they experience emotions such as fear, anxiety and happiness. They share the intellectual capacity to create and use tools, learn and teach other languages. They remember their past and plan for their future. It is in recognition of these and other morally significant qualities that the Great Ape Project was founded.

This philosophy has been put into practice with a vengeance in 2007 in both Spain and Austria. In Austria it has been used to attempt to get the courts to grant a chimpanzee, Hiasl, human rights. The *Guardian* newspaper published its story on Hiasl, the 'Austrian' chimp, on April Fool's Day. Unfortunately it was no joke. Science had given way to something approaching legal lunacy.

Hiasl had arrived in Austria in 1982, smuggled in from Sierra Leone as fodder for biomedical research. Customs officers intervened and he ended up being raised in a human family until he was 10 years old, before being housed in an animal sanctuary. It was the closure of the sanctuary that precipitated the legal clamour for rights for Hiasl, now 26 years old, because sale of Hiasl to a zoo or pharmaceutical company (unlike the UK, chimps can be used for medical research in Austria) would have helped pay off the sanctuary's debts. An Austrian businessman donated 5,000 euros—roughly equivalent to a month's food supply and veterinary bills—jointly to Hiasl and Martin Balluch, president of the welfare group Association Against Animal Factories, with the odd proviso that both man and chimp had to agree on how the money would be spent. Unfortunately, under Austrian law, only a human can receive a personal donation. It would seem sensible to have created a little foundation to administer this cash, and any other donations to Hiasl's welfare the seed money prompts in the future, but, instead, it has prompted Balluch to seek a legal guardian for Hiasl, now re-christened Matthew Hiasl Pan, and, since only a human can be granted a legal guardian under Austrian law, Balluch's request, essentially, is that Hiasl be accorded human rights. The proposed legal guardian is a British woman, Paula Stibbe.

Much has been made in the popular press of Hiasl's 'humanoid' behaviour. He loves hide-and-seek games, likes being tickled, watches TV, paints pictures, and recognizes himself in a mirror. 'Our main argument is that Hiasl is a person and has basic legal rights,' said Eberhart Theuer, the lawyer representing the group. 'We mean the right to life, the right to not be tortured, the right to freedom under certain conditions,' Theuer said. 'We're not talking about the right to vote here.'

Two world-famous primatologists, Jane Goodall and Volker Sommer, backed Hiasl's case. Sommer's brief to the court said 'It is untenable to talk of dividing humans and humanoid apes because there are no clear-cut criteria—neither biological, nor mental, nor social'. Both

Goodall and Sommer trotted out the by now standard argument that chimps and humans are nearly 99% genetically similar.

Not all biologists found the 'scientific' argument in the Hiasl case convincing. The BBC reported Professor Steve Jones, of University College, London, retorting:

> As most people know, chimpanzees share about 98% of our DNA, but bananas share about 50%, and we are not 98% chimp or 50% banana, we are entirely human and unique in that respect. It is simply a mistake to use an entirely human construct, which is rights, and apply it to an animal, which is not human. Rights come with responsibility and I've never seen a chimp fined for stealing a plate of bananas!

Eventually, the judge in the case, Barbara Bartl, ruled that Hiasl was neither mentally handicapped nor in imminent danger (the two main criteria under Austrian law for appointing a legal guardian) and his legal representatives are now busy planning an appeal.

What is happening in Austria is part of a trend. In 1999, New Zealand had granted all species of great apes rights as 'non-human hominids'. The Great Ape Project today is behind a more wide-ranging attempt to enshrine human rights for apes into Spanish law. They were also responsible for the New Zealand law. The proposed Spanish law goes beyond granting apes the right to life, protection of individual liberty, and prohibition of torture, to include banning private ownership of apes or their use in entertainment. It would also, if passed in its present form, oblige the Spanish government to convene an international forum on the issue of granting rights to apes world-wide. In mid-2008 this recommendation was passed by the Spanish parliament.

The Green MPs supporting the motion in the Spanish parliament use stock phrases such as 'so close to humans', and 'they show a degree of intelligence and awareness and, indeed, self-awareness', and 'their social and emotional needs are at the same level as handicapped people, small children, elderly, mentally impaired people—and they all have rights'.

I personally find it quite distasteful to see chimpanzees described as equivalent to tiny children or the mentally impaired simply because, while both latter categories have human rights, they need to be cared for due to either their immaturity or mental frailty. It suggests that chimpanzees are mentally sub-normal human beings, which I find insulting to both species. Claims for genetic and cognitive proximity of chimpanzees to humans are used time and time again to bolster this equivalence. Even if chimpanzees are 99% genetically similar to human beings, that does not make them humans, and it does not make them persons—the only category which, by law, can be granted human rights. Rights, as Steve Jones points out, have nothing at all to do with genetics; they are a human social construct.

Sometimes, amid all this scientific talk of genetic and cognitive similarity we can lose sight of the most important facts that separate the two species. It is humans that have speech and language, humans that have culture, art, music, science and technology, humans who remember the past, plan for the future, fear death, and pay taxes. Simplistic commentators who evoke the 98.5% genetic similarity between chimps and humans invite us to think this means humans are 98.5% chimps and chimps are 98.5% human. In the days before any type of protein or DNA comparison was possible, when we only had palaeontology and African fossils with which to chart human origins, chimpanzees were not part of the picture. Only one fossilized find claiming to be that of a chimpanzee has ever been made—and that quite recently. Humanity began after the split from the common ancestor with the Australopithecines and then the genus *Homo*. The need to compare living DNA with living DNA has forced us to extend our embrace to chimpanzees and we sometimes forget that we are not descended from them, but something that might have looked and behaved quite like them, or equally well might not.

Some commentators on human evolution doubt very much whether we will find the answer to human origins in genes. I disagree. I believe we *will* find much of what we are looking for in the comparison of genomes, and that genetic differences between us and chimpanzees, and the other great apes, *properly interpreted*, are already providing us with a number of tantalizing clues. The job would be easier if the three comparative sciences involved were better integrated, but, all too often, the links between genetics, cognition, and behaviour, and the functional structure of our brains and nervous systems, have not yet been properly made. The job would also be easier if scientists, conservationists, and philosophers refrained from conflating species comparisons between chimps and humans with the need to protect chimps and the other great apes because they are 'nearly men'. I want to take a hard look at recent science to see if it really supports the 'chimps are us' argument, and suggests that the differences between humans and chimps are 'merely' quantitative. Are we simply remodelled apes or unique primates after all? This is science in the making: we are trying to hit a moving target—and that target is moving rapidly. Today these inroads are increasingly being made using powerful genome technology to sequence and compare vast tracts of chimp and human genomes. This tends to overshadow research that has already stumbled over tantalizing clues to human cognitive evolution as a by-product of work in psychiatric genetics. These were among the first proposed 'genes that made us human' and it is with a few examples of this serendipity that I want to start my own exploration of chimp–human differences.

CHAPTER 2

The Language Gene That Wasn't

KE family child. 'Looh ... ..ah'
Interviewer. 'And where do you live, Laura?'
Child. Unintelligible.
Interviewer. 'And how old are you?'
Child. 'Foa. Foa wes.'

O ne of the most extraordinary scientific detective stories of modern times concerns the discovery of a gene for speech and language impediment in a large British family. It is a tale that is full of serendipity, riven with scientific controversy, and with important implications for human evolution.

Specific Language Impairment (SLI) tends to run in families but even by the standards of SLI the KE family are an exceptional example. Elizabeth Auger's special educational needs unit at a primary school in Brentford, west London, was inundated with them. At one point, in the late 1980s, she had seven children in her unit at the same time, all from this one extended family. Several family members from the older generations had been previously referred to her. They all suffered, to a greater or lesser extent, from a kind of palsied speech. It was as if the lower part of their faces was somewhat frozen, they simply couldn't form words properly. Several of them had halting speech, as if they were searching for words. Sentences were short and broken and vocabulary was limited. Their speech has been called telegrammatic. When you watch videos of the children struggling with even the most simple sentences you see a mouth that simply will not go where it needs to go to form fluent words. Consonants were a living nightmare. They were omitted, elided, or approximated, as in words like 'boon' for 'spoon', 'able' for 'table', and 'bu' for 'blue'. Polysyllabic words like 'parallelogram' would be completely beyond them. They would often neglect to inflect verbs, using only the present-tense stem. Even the more fluent, older, affected members of the family only managed some fluency of speech by gliding over the normal transitions that clearly mark individual syllables, and by ignoring distinctions between singular and plural, or appropriate verb tense endings, and by filling up the awkward gaps with shrugs, hand gestures, and grunts.

Auger decided it was high time this family was properly investigated and raised the matter with a medical geneticist, Michael Baraitser, of the Institute of Child Health in London. Baraitser discussed it with his head of department, Marcus Pembrey, and Elizabeth Auger was invited to a meeting at the Institute, along with two affected members of the family. Most linguists at the time were very hostile to the idea that genetics could be involved in language impairment, but to Baraitser and Pembrey, the occurrence of such similar symptoms across such a large family suggested the opposite. However, with such a complex phenomenon as language, it would be naive to imagine that those genetics would be a simple matter: it would be far more likely that many genes were involved. Nevertheless it was very exciting and their registrar, Jane Hurst, was assigned to visit the family, note their symptoms, and take blood samples for genetic testing.

Hurst, although a clinical geneticist and not an expert on the psychology of language, found no difficulty distinguishing between affected and unaffected members of the family. She used no tests for language ability whatsoever, she just listened to them talking. In all, throughout three generations, she counted 16 affected individuals. No unaffected parent had ever produced an affected child, and every affected parent had passed on the language disorder. Painstakingly, she drew the first family pedigree for the KEs and it quite clearly showed something that none of the researchers had expected. This particular language impairment did not require an orchestra of genes at all, it bore the tell-tale pattern of the effect of one single dominant gene, on one of the chromosomes other than the sex chromosomes.

Gene mutations are described as being either dominant or recessive. If a gene mutation is recessive we must receive two copies of it, one from our mother and one from our father, in order for it to lead to a trait or characteristic that shows up in us. Scientists call this the homozygous condition. However, if a gene mutation is dominant, its effect can never be masked even if we receive only one mutant copy from either parent. Each child therefore has a 50% chance of inheriting that gene mutation, and the trait that goes with it. This was the pattern noted in the KE family. Could this be the first evidence of genes for language? The BBC was onto it in a flash. Could they film the bloods being sampled? Hurst remembers that only one blood sample remained to be taken, that of the all-important grandmother, the oldest surviving affected member of the family, from whom all impairment flowed. Well into her first pregnancy, Hurst recalls the difficult car journey from her home in Brighton to west London in January 1989. 'I felt terrible with morning sickness. I thought, "Oh! God! I'm going to be sick on camera, taking her blood!"'

Meanwhile, over in Montreal, Canada, the psycho-linguist Myrna Gopnik was preparing to visit her son and his family in Oxford, England. Gopnik had recently founded a Cognitive Science Group at McGill University which met regularly and included paediatricians. One of them had introduced her to an interesting 14-year-old patient who had great difficulty constructing fluent narrative speech. He spoke in very simple sentences and failed to use proper tenses and plurals, where appropriate, for words. His father shared the same difficulties. Although a successful computer scientist, he had no simple everyday conversation and his speech was laboured and agrammatic. It reminded her of the speech of aphasics, whose language difficulties arise in adulthood because of brain traumas like strokes, especially if they affect Broca's Area, the main language processing centre of the brain. While she was talking to the father she noticed he had very poor inflection. He told her he had 'two computer' for instance: he couldn't make the plural. His wife told her she thought her husband's language was impaired and that it took great effort for him to construct simple sentences with inflected words in them. Even simple conversational sentences like 'We built the pool about five years ago' required great and laboured thought. It reminded Gopnik of how tired one gets after a day of stumbling through conversation abroad in a foreign language. These people behaved, she thought, as if they had no native language. None of it 'came naturally', it all had to be computed.

Some years earlier, Noam Chomsky had shaken the world of linguistics by pointing out something that Darwin had previously drawn attention to: that children pick up languages astonishingly easily, yet, of course, they do not automatically seem to know how to bake cakes. Chomsky had proposed that children learn language too fast to be absorbing the rules of grammar via exposure to speech in everyday life. Their brains must be pre-disposed to it. There must be a construct, he said, called Universal Grammar already programmed into young brains. Chomsky fell short of any accurate biological description of his language organ and never mentioned genes. However, Steven Pinker, formerly a colleague of Gopnik's at McGill, but then at the Massachusetts Institute of Technology, was expanding Chomsky's theory into a truly biological dimension. He suggested that language was constructed in the brain by a number of computational modules, each responsible for a particular component, like the appropriate inflection of words. It was likely, said Pinker, that each of these computational modules would be underpinned by its own discrete genetic foundations.

While in Oxford, Myrna Gopnik was invited to give a guest lecture at the Psychology Department. Wandering around the building with her daughter-in-law, her gaze alighted on a bulletin board which gave notice of interesting forthcoming events. The BBC's *Antenna* science television series, it noted, was transmitting a film that week about a family in west London who appeared to be carrying a gene for language disorder. This was the very film for which the heavily pregnant Jane Hurst had forced herself up from the south coast. A few days later the Gopnik family watched, fascinated. Could this be one of the genes Pinker had invoked?

Gopnik located the primary school, rang Elizabeth Auger, and obtained an invitation to visit and give an informal talk about language impairment, based on her observations of language-impaired individuals in Montreal. In the audience was the KE grandmother and several other family members. Much of what Gopnik had to say resonated with her audience and she was invited by Auger, and encouraged by the grandmother, Mrs. K, to study them. Gopnik recalls being told by Auger that, although she had mentioned the KE family to members of the Institute of Child Health, there had appeared to have been little interest, leaving the field open. This, although Jane Hurst had been taking bloods since 1987! Gopnik agreed to work with the KEs and returned to Montreal to prepare a battery of tests.

By now, the Institute of Child Health's own language expert, Faraneh Vargha-Khadem, was also on the case and had already begun testing family members in her language laboratory. Why was Gopnik invited on board at all? It may be that Mrs. K was deeply resentful of some of the comments made about her family's impairment. Conventional linguists were anxious to suggest an environmental route to the disorder and there had even been a peculiarly British social slur analysis which had suggested that, since the family were working class, mangled words and poor vocabulary were to be expected. For Mrs. K, having the family at the centre of important scientific investigations was an excellent way to regain lost dignity. Scientists? For once, you couldn't get enough of them.

The problem for the British scientists was not one of lack of interest but lack of cash. Marcus Pembrey, like many of his British contemporaries then, as now, was a pauper king. Although he had been fortunate enough to have stumbled over such a large family pedigree, and had taken advantage of that by collecting a large number of blood samples, there was no money in the kitty to find the genetic culprit. Pembrey decided to try to get Gopnik on board by inviting her to pool her results with those of Vargha-Khadem and the others. When Gopnik resisted, the Dean of his Institute questioned Gopnik's access to the family by suggesting to his opposite number at McGill that her intrusion would cause unnecessary stress on this vulnerable family and had not received ethical approval. His opposite number declined to intervene.

Gopnik waded in with a battery of tests originally designed for the study of aphasia. This has always been criticized because aphasia, as we have already seen, is strictly a language disorder caused by damage to the language processing centre of the brain in later life. The KE family clearly inherit their disorder and so it is present from birth. Furthermore, whatever the language problems, they were accompanied by severe difficulties in the production of words. Their disability appeared to involve speech *and* language. Nevertheless, Gopnik concluded that the core deficit in the KE family concerned an inability to change the tenses of words and the lack of a general rule for producing appropriate plurals. For instance, they might be able to distinguish between a picture of a solitary book, versus a pile of books, when asked to point to 'the book', a known word they might have filed in memory, but when shown a picture of an imaginary animal called a 'wug' they could not describe a group of these animals as 'wugs'. Altogether, she said, they lacked the underlying grammatical rules for changing word endings to indicate singular, plural, or tense. They could therefore not do it automatically, but only if those words had been stored in memory. The defect, she stated, was not one of general cognition; some aspects of grammar were spared, and it was restricted to one feature of grammar. Astonishingly, she made no mention of the severe facial palsy of the KE family, a palsy others had long noted to be so severe as to prevent them articulating the sounds of speech. She had turned a blind eye to their inarticulacy.

Gopnik rushed into print, and published a letter to *Nature* in April 1990 titled 'Feature-blind grammar and dysphasia'. (Dysphasia is a mild form of aphasia.) The British scientists had implored her not to publish and warned her that she did not have the full story. When she went ahead anyway they accused her of simply 'gathering the low hanging fruit'. To this day, Gopnik asserts that, since the same grammatical faults turned up in members of the KE family when asked to write, the core deficit must be grammatical, and not one of articulation. Furthermore, she asserts, *she* never had any real problems understanding what they were saying, even on the telephone. For the next seven years she stood her ground, arguing strenuously that the articulation difficulties were a distraction, and implying that the London-based scientists, by stressing inarticulacy, were arguing that there *were* no underlying grammatical difficulties at all.

In the same year that Gopnik published in *Nature*, Steven Pinker published a major paper in *Science*, titled 'Rules of Language'. Pinker's battle was with those psychologists who continued to maintain that the human mind was a general-purpose learning mechanism—a homogeneous computer. He selected one aspect of language and provided evidence that the mind has a discrete module that operates a grammatical rule to form the tenses for regular verbs. This inflection rule turned 'walk' into 'walked' and 'turn' into 'turned', whereas the tenses of irregular verbs ('sit-sat', 'feel-felt', and 'tell-told', for instance) were generated from memory. He concluded:

> Focusing on a single rule of grammar, we find evidence for a system that is modular . . . more sophisticated than the kinds of rules that are explicitly taught, developing on a schedule not timed by environmental input, organised by principles that could not have been learned, possibly with a distinct neural substrate and genetic basis.

In her paper, Gopnik had referred to the evidence Marcus Pembrey's group had unearthed for a single gene responsible for the language difficulties experienced by the KE family. Pinker had posited a brain module responsible for a single feature of grammar, Gopnik had discovered it in a single large family pedigree, and it had a simple genetic basis. QED.

In his book *The Language Instinct* Pinker records the press fanfare in 1992 when Gopnik presented a paper at the annual meeting of the American Association for the Advancement of Science. 'Better grammar through genetics' ran one headline, while another quipped 'Poor grammar? It are all in the genes!'. Yet, hidden away in a much more obscure journal, in 1990, a paper by Jane Hurst, Michael Baraitser, and the staff at the special school in west London painted a very different picture of the KE family's impairment. Entitled 'An extended family with a dominantly inherited speech disorder' (notice the stress on *speech* rather than *language*), the paper reported that the family had serious communication difficulties due to a severe verbal dyspraxia, a lack of control over the facial muscles that allow the mouth to articulate the sounds of speech. There were also language difficulties: simple speech and poor comprehension. For instance, 'the girl is chased by the horse' became 'the girl is chasing the horse'. They concluded that the core deficit in the KEs involved *both* speech and language. So what was really going on in the KEs? Did they suffer from a language impairment, an articulation impairment, or both?

This acrimonious scientific disagreement took the next ten years to resolve. In fact it wasn't until 1995 that Faraneh Vargha-Khadem was even ready to publish her exhaustive description of the KE family trait. It suggested that Gopnik's account of what was afflicting the KE family was far from the whole story.

Although Vargha-Khadem agreed with Gopnik that the affected KEs were unable to inflect verbs with appropriate tense endings, she showed that this inability extended to irregular verbs and was not, as Gopnik had claimed, restricted to regular verbs. However, she also showed that there was an increased occurrence of over-regularizations in verb endings, like 'heared, finded, goed'. These 'tense mistakes', as any parent will tell you, are typical of very young children at the outset of language acquisition, suggesting that the KE family possessed at least some part of the grammatical rule for marking tense.

But comprehending language and constructing language in the brain is only half the problem. To actually *produce* the sounds of language involves a huge amount of fine motor control of the lips, mouth, tongue, and pharynx; the whole vocal tract. The tongue alone makes up to a hundred movements a second when you are talking. The mouth and lips have to coordinate as well. Vargha-Khadem found that the affected members of the KE family were impaired on this motor control, whether it involved attempts to string sentences together or even to make a string of facial movements like 'close your lips, then open your mouth, then stick out your tongue'. Whatever had gone wrong with the KE family involved language, speech, and mouth and facial neuromuscular control. It was a very broad or diffuse trait, or phenotype, and provided, she reported in a calculated swipe at Gopnik, 'no support for the existence of "grammar genes"'.

Were the KE family's problems with speech and language mirrored by abnormalities in the structure of their brains, and the way they worked? Vargha-Khadem and her colleagues used two imaging techniques to test this: PET (positron emission tomography) and MRI (magnetic resonance interferometry). PET scans tell you which parts of the brain are being used on a particular task and how hard they are working by, in effect, measuring the rate at which glucose is being burned as fuel for energy. MRI measures the size, shape, and density of particular parts of the brain, giving an anatomical, as opposed to a metabolic, picture. Using these techniques, Faraneh Vargha-Khadem and her colleagues have shown that the brains of affected members of the KE family function very differently to those of normal people. There are clear abnormalities in parts of the brain involved in both motor actions and speech. Of particular interest was the caudate nucleus, a tail-like organ which is part of the basal ganglia, deep in the brain. The basal ganglia are important for fine, coordinated muscular control, exactly what is needed when our vocal tracts make bewilderingly fast movements as we speak. The caudate contained less grey matter in affected family members than in normal subjects and, gratifyingly, its volume correlated with their performance on tests of coordination of mouth movements, and tests of nonsense word repetition which measure the brain's ability to process arbitrary words and re-assemble the phonemes involved so that they can be pushed back out again as speech. Here, memory cannot be involved because the word is both novel and nonsense.

The PET scans suggested the caudate was not only smaller but was having to work harder in the affected KE family members, especially when they had to repeat words. Another basal ganglia structure, the putamen, had significantly more grey matter than in controls. Other patients with combined pathology of these two areas are unable to properly pronounce words and syllables, and find it difficult to position the face, lips, and tongue to make speech sounds. Elsewhere, the cerebellum (a brain organ at the rear of the head, just above the brainstem), known to be involved in speech articulation, was anatomically abnormal, especially in regions of it that are involved in word generation. There was also abnormality in the planum temporale, a small brain region in the cerebral cortex which is slightly larger on the left than on the right, and is related to verbal ability.

Although there are clear grammatical abnormalities in the affected members of the KE family, there are only three tests for which the affected members of the family *never* overlap in their scores with non-affected members. These are the tests for complicated, sequential, facial movements ('first open your mouth wide, then close your lips tightly together, then make the "ah" sound'), and for word and non-word repetition. Thus higher regions of the brain were affected, most pointedly Broca's Area, which does the higher-order language processing, and the basal ganglia structures Broca's Area is connected to that control the complicated muscular orchestra we call speech. In this way both speech and language were impaired by the mutation of just one gene.

Vargha-Khadem and her colleagues had painstakingly built up a detailed picture of a complex condition that involved speech, language, and neuromuscular control. Their clear portrait of structural and functional brain abnormality involved many of the parts implicated in aspects of grammar *and* language *and* articulation. For the first time, in 1998, Myrna Gopnik changed her tune. In a remarkably conciliatory paper she acknowledged the existence and relevance of Hurst's original observations on the KE family dyspraxia, and Vargha-Khadem's vast corpus of neuroscience. She really had no choice. Vargha-Khadem had exhaustively described significant brain pathology for something Gopnik had chosen to ignore, the palsied speech, and shown it to be a central feature of the KE family disorder. Soon afterwards Gopnik slipped out of the picture and into retirement and simplistic talk of 'grammar genes' went with her.

This left the London-based scientists still no nearer to finding the single gene ramifying through the KE family that *was* responsible for their grammatical defects, poor speech articulation, and lack of facial muscular control. What type of gene could be responsible for all this structural and functional abnormality? Faraneh Vargha-Khadem was getting impatient with Marcus Pembrey. She had spent five years pinning down the phenotype of the KE family disorder in excruciating detail. She was fearful that another scientific group would sweep in and grab the genetics.

Pembrey's money-juggling had finally run out of steam, and he no longer had the funds to move the hunt for the gene forward. A powerful ally was needed. They approached Tony Monaco at the Wellcome Trust Centre for Human Genetics, in Oxford. Monaco was eager to help. A new post-doctoral researcher, Simon Fisher, was put in charge of the investigation. Simon recalls a lazy summer holiday spread-eagled on a beach reading Steven Pinker's book *The Language Instinct*. He was fascinated by Pinker's account of the KE family, Gopnik's work, and the evidence for the surprisingly simple genetics at the root of the disorder. 'I thought, "I'd give my eye-teeth to be able to research something as interesting as that", and when I turned up for work at the Wellcome—there it was—waiting for me!'

The search for the gene accelerated. Using a powerful new battery of genetic markers they narrowed the gene's location down to a section of the long arm of chromosome 7 containing some 6 million base-pairs of DNA. A great leap forward but still tantamount to a Chief of Police announcing 'We're closing in on the culprit, he's holed up in a house somewhere in Greater London!' In a constant reminder to the scientific community of the provenance of the gene they gave it the name SPCH 1. By consulting online databases of known DNA sequences they knew this region, code-named 7q31, contained at least 70 genes. There was nothing else they could do but painstakingly screen the whole area looking for a tell-tale mutation. This long march down chromosome 7 was then cut short by a spectacular piece of luck as Jane Hurst once again entered the picture.

As a clinical geneticist, Hurst had become involved in the case of a pregnant woman who had undergone a routine amniocentesis to test for Down's Syndrome. When the genetics laboratory examined cells from her baby, withdrawn from the amniotic fluid, they discovered that a translocation had occurred on chromosome 7. Translocations occur between non-homologous chromosomes (chromosomes that are not one of a pair). Occasionally, bits of one chromosome can become detached and then re-attach themselves by mistake to another. In this way, the DNA on the detached piece of chromosome becomes translocated to somewhere else in the genome. In the case of this baby, CS, the exact point of the break was known. A piece of chromosome 7 had broken off and re-attached itself to chromosome 5 with a piece of chromosome 5 going in the opposite direction. At the time this was discovered the clinical genetics team had no idea whether or not this translocation would cause problems for CS. They kept a watching brief.

Although CS appeared to develop normally as a baby, it was noticed in 1998, when he was about two years old, that his language was delayed and that he had speech articulation difficulties remarkably similar to those of the KE family. He was referred straight back to Jane Hurst at about the time Simon Fisher, Faraneh Vargha-Khadem, and colleagues reported narrowing down the search for the KE family gene to the region of chromosome 7 at 7q31. Hurst noticed that CS's translocation had occurred right inside the region the Wellcome team had discovered. She immediately picked up the phone to Fisher, 200 yards down the road, and told him: 'I've got the boy who is going to get you your gene.'

And so it proved. When they narrowed the search down to the exact point at which the chromosome breakage in CS had occurred they discovered that it had disrupted a gene for which they could find no previous record. They cross-matched with the KE family and found the same mutated gene in those whose speech and language were affected. The gene was given the name FOXP2. It was a new member of an ancient family of genes that have been steering embryonic development from fungi to mammals for millennia. Careful step-by-step analysis of the DNA sequence of this gene from the KE family, by graduate student Cecilia Lai, revealed one single point mutation, one single DNA nucleotide change—small, but enough to cause cognitive havoc. FOXP2 was certainly not a 'grammar gene' as such, but turned out to play a far more fundamental role in the chain of genetic events that lead to language.

Faraneh Vargha-Khadem has a neat party trick she likes to play on visitors who come to talk to her about her research, a trick that illustrates the very profound role of FOXP2. She plays them videotapes of the tortured speech of the KE family and then follows that up with a clip of an interview with a teenager called Alex. As Alex recounts a Lonely Planet-type backpacking holiday around Europe it gradually dawns on you that, while his powers of description are adequate and his use of English reasonable, his speech more resembles that of an 8 or 9 year old, than someone who is clearly at least 18. The reason? Alex had severe brain disease when he was very young and at age 9 he underwent a complete hemispherectomy: the entire left side of his brain had to be removed. He had never spoken in his life. After the operation he learned to speak even though his brain had to make use of structures in the right hemisphere to replace the lost areas normally used for speech and language in the left, and even though the critical period in brain development for language acquisition, the 'window of opportunity' during early childhood, had long been passed. On the one hand, the human brain is so awesomely plastic and adaptable that even radical surgery fails to de-rail speech and language, yet a tiny point mutation in a single gene, FOXP2, leaves affected individuals with dramatic impairments in speech articulation and language comprehension and production from which they never recover, despite decades of speech therapy.

The reason for the widespread effects of FOXP2 lies in the fact that it is a master-controller gene, one of a number of genes that produce transcription factors. These are proteins that bind to regions of the DNA of a number of downstream target genes to regulate them—turn them on or off. The mutation in the KE family occurred in a region of the gene that codes for a short but crucial section, only 80 to 100 amino-acids long, of one of these regulatory proteins. FOXP2 controls the activity of a genetic orchestra that, in some way, sculpts the neural pathways of the brain for the purpose of language and speech articulation. In an elegant piece of research, which finally nailed down the FOXP2 story, it has been shown that the sites in the brain where FOXP2 is normally active agree closely with the sites of brain pathology noted by Faraneh Vargha-Khadem and her team, using brain-scanners. Disruption of FOXP2 in these areas was associated with the brain pathology that caused the KE family disorder. This was the real QED!

Soon after the discovery of FOXP2, Tony Monaco, at the Wellcome Institute, had an important visitor. Svante Paabo is head of the Molecular Genetics Department at the Max Planck Institute for Evolutionary Anthropology in Leipzig. He was the first scientist to clone Neanderthal DNA, some 35,000 years old, from tissue retrieved from the long bones of a skeleton from a German cave. Paabo is a world leader in the attempt to single out the genes that make us human and one of his key strategies is to comb the world of psychiatric and psychological genetics for interesting mutations. Paabo had arrived on a fishing expedition. Monaco had FOXP2 to offer and they agreed that Paabo's laboratory should look to see if it had an interesting evolutionary tale to tell. Paabo entrusted the work to his research lieutenant Wolfgang Enard, who compared the amino-acid sequence of the FOXP2 protein in mouse, rhesus macaque, orang-utan, gorilla, chimpanzee, and human. He discovered that it had hardly changed at all over the 130 million years of evolution that separate the common ancestor of chimps and humans from the common ancestor that led to the mouse. Chimp, gorilla, and macaque sequences were identical and only one amino-acid change separated them from the mouse. However, a further two amino-acid substitutions have occurred over approximately 6 million years in the hominoid line that led to us. The pace of evolution had suddenly picked up. They looked at patterns of variation in the gene throughout a number of contemporary human groups from Africa, Europe, New Guinea, Asia, and South America, and their mathematical analysis suggested that these two amino-acid changes had been the target of natural selection. The two mutations had occurred and become fixed in human populations probably as recently as 200,000 years ago, the approximate date when modern humans, *Homo sapiens sapiens*, arrived on the scene.

The evolutionary tale of *FOXP2* dovetails nicely with the evolution of our human lineage. But it gets us no nearer a precise description of exactly how *FOXP2* is implicated in speech and language, grammar and articulation. The exact nature of the two amino-acid substitutions unfortunately tells us very little. Simon Fisher speculates that it may have caused a subtle change in the regulatory protein that decreased its ability to switch other genes on or off. Puzzlingly, one of the two 'unique' human lineage amino-acid substitutions also turns up in the great cats, and, with the exception of the Chronicles of Narnia, talking lions are thin on the ground. How the one surviving 'unique' amino-acid substitution can give rise to speech and language, something the great apes so conspicuously lack, remains, for the moment, a mystery. Since it is obviously unethical to excise the gene from humans and see what happens, Enard and Fisher are collaborating on a series of experiments with 'knock-in' mice where the human variant of the *FOXP2* gene is incorporated into the mouse genome and they can see what changes, if any, occur. Precisely what changes might be expected is not clear. 'We are not expecting the mice to tell us,' jokes Enard.

In the meantime, some exciting new information on *FOXP2* has arrived, thanks to studies of convergent evolution in birds and humans. (Convergent evolution occurs when evolution solves a problem the same way in two entirely unrelated species. For instance, the eye has arisen independently many times throughout the animal kingdom.) Bird-song has many similarities to human language: the basic 'vocabulary' of a bird-song is learned by imitating older family members, there is a critical period in development when young birds can acquire it, and it is plastic in that basic patterns of song are honed by each bird to personal taste. It turns out that the expression of FOXP2 in bird brains is particularly strong in those parts of the brain involved in song, areas which are comparable to the parts of the brain in which these two key genes are expressed in humans. The evidence has come from Stephanie White, in the United States, and Constance Scharff, in Germany, who have been working with zebra finches and canaries.

Zebra finches are fully mature by 90 days old, but the young male birds (only male zebra finches sing) have already acquired their model song by 40 days. There is a sensitive period for song acquisition between about 20 and 35 days during which the young finches start off, like human babies, by babbling, and proceed to more coherent song. They are tutored by their fathers, other older- generation males, and, occasionally, older siblings, and first achieve a copy, or rendition, of the tutor's song. However, this basic framework is then individualized by adding notes and making other changes.

Scharff and her colleagues discovered that, while FOXP2 is expressed in a broad range of brain structures, its expression is particularly intense in a part of the basal ganglia called Area X, which is essential for vocal learning. This heightened activity begins when the chicks are between 35 and 50 days old, just as vocal learning occurs, and continues throughout life. In canaries, song type is seasonal. For parts of the year canary-song is very stereotyped, but at other times the song becomes unstable and plastic; its owner is experimenting. During these creative periods FOXP2 expression soars in Area X. The two research groups also recorded strong FOXP2 activity in a variety of structures that connect the basal ganglia (the location of Area X) with areas responsible for hearing and controlling the throat movements associated with bird-song.

Just as in the human work, these researchers have found that FOXP2 has a wide role to play in developing bird and human brains, but a specialist, more prominent, role in those parts of the brain that coordinate hearing and singing/talking. They conclude that these genes create a 'permissive environment' in parts of the brain within which vocal learning can evolve if, and once, other factors come into play.

Information from birds, together with Faraneh Vargha-Khadem's imaging studies on the affected members of the KE family, show us that the articulation of speech and the comprehension of speech use the same circuits in the brain. Language is not abstract, but grounded in and restrained by the physical process of articulation. Words are what our mouths can make them. As Vargha-Khadem puts it, language cannot be free-floating in the brain. As a comparison, she says, all humans have the inherent ability to walk, but it cannot be realized without the feet and legs to go with it, and, more precisely, the sensorimotor systems (those neural systems that control muscular responses to sensory information) in the brain that make them work. In the same way, language must have the appropriate sensorimotor system to map onto—the right mechanism in the brain with which it can fuse.

In a recent scientific paper, commenting on the role of FOXP2 in speech and language, the celebrated linguistic scientist Philip Lieberman makes the telling point that Broca's aphasia never occurs thanks to damage to Broca's Area, in the cortex, alone. It requires damage to the basal ganglia as well before the classic speech and language deficits show up. This failure to mention the role of these sub-cortical 'primitive' brain structures, Lieberman argues, goes right back to Broca himself in 1861, when he failed to point out that the key patient on whom his description of Broca's aphasia was based had extensive damage to basal ganglia structures as well as the eponymous cortical region. It was rather like praising the captain on the bridge for a successful ocean crossing and forgetting the sweaty endeavours in the engine room.

Broca's modular view of the 'seat of language' being situated in the higher-order 'advanced' cortex was subsequently followed by

Chomsky and Pinker. More modern brain research, says Lieberman, has shown that circuits looping out of the basal ganglia, into the cortex, and back again, are involved in the fine coordination of muscular actions that enable us to walk, dance, and contort our faces. They also allow us to make novel combinations of a finite repertoire of muscular actions of the vocal tract to form an infinite variety of words, and allow our brains to make an infinite number of novel sentences out of a finite number of words, using a finite number of syntactical rules. To imagine what he means, remind yourself of how many words fill the New Oxford Dictionary, all comprised of different mixtures of the twenty-six letters of the alphabet, or how the huge number of different proteins in the human body are all made by different combinations of only four nucleotide molecules in the DNA double helix.

Lieberman calls the basal ganglia a 'sequencing engine' deep in our brains, which can operate in all these contexts. Lieberman doesn't dispute that the content of thought, memory, and ideas are higher-order functions of the brain, but reminds us that it is these older brain structures that make speech, syntax, and creative thought possible. This is the 'permissive environment' evoked by those scientists studying bird-song, and we need to know how it is that subtle DNA sequence changes in human *FOXP2* have led to crucial changes here to give us something apes so palpably lack, the power of speech and language.

Not surprisingly, several groups world-wide are now trying to discover the array of 'downstream' genes that seem to be under the control of *FOXP2* in the brain. Although *FOXP2* is not directly implicated in any other type of language disability, an international group of scientists, drawn mainly from the Wellcome Centre for Human Genetics in Oxford, has discovered one member of this 'downstream orchestra', a gene called *CNTNAP2*, whose activity is under direct control by *FOXP2*. *CNTNAP2* is active in the developing cortex of the brain and variants of the gene have been associated with specific language impairment and language delays in children with autism. Two recent pieces of research bolster the twin ideas of FOXP2 involvement in language as a form of social communication, and Lieberman's idea that it is active in a 'sequencing engine' in the brain, responsible for rapid articulation of sounds.

In 2005, a group of scientists led by Joseph Buxbaum, of the Mount Sinai School of Medicine in New York, did the opposite of 'knocking in' the human FOXP2 gene to mice. They 'knocked out' murine FOXP2. When both copies of the gene were tampered with the mice were very severely affected, their movement was badly impaired, and they died prematurely. However, when only one copy of FOXP2 was disrupted there was only mild developmental delay in the cerebellum and they were left with a communication block. When lifted away from their mothers, the pups failed to make any ultrasonic distress calls. When Buxbaum's team analysed the vocalization patterns and the bandwidths of the vocalizations they concluded that much of the machinery in the vocal tract and brainstem associated with vocalizing was normal. The problem lay with subtle changes in the cerebellum. This makes complete sense because the formation of speech sounds is a motor activity and the cerebellum integrates this fine motor control. The cerebellum was one of the brain areas most affected in the KE family.

In late 2007, a group of Chinese scientists together with Stephen Rossiter, from Queen Mary College, London, reported on some fascinating research that connected mutations in the FOXP2 gene to echolocation in bats. Received wisdom, they said, was that FOXP2 was extremely conserved throughout the animal kingdom—there was no sequence variation—until you get to the comparison between humans and chimpanzees, where, as we know, there are two amino-acid substitutions in humans. This has led, they argue, to exclusive consideration of the role of FOXP2 in the context of the emergence of language—which humans have and chimps don't. However, when you compare echolocating with non-echolocating bats, they report, you find a range of sequence differences in *FOXP2*, suggesting that the correct context to explore the action of *FOXP2* is sensorimotor coordination, not language per se—exactly the same conclusions reached by Varga-Khadem.

Echolocation is an extremely sophisticated business, being used for orientation and prey capture at high speeds. Many insects use very unpredictable, almost chaotic, flight patterns called protean behaviour, designed to lead predators a merry dance. To prevent themselves banging into things, and to fill their bellies, bats emit echolocation pulses at up to 200 sounds per second, decipher the bounce-back in milliseconds, and make necessary motor adjustments to their flight controls within the same tiny time-frame. Receiving and transmitting this ultrasonic barrage of sonar information requires an extraordinary and complex coordination of hearing, and nasal and orofacial movements, since the ultrasonic signals bats use for navigation are formed in the larynx. Bats, like song-birds and humans, also exhibit vocal learning. It makes language look like child's play.

When the scientists sequenced the entire *FOXP2* gene in a range of echolocating and non-echolocating bats they found many different amino-acid substitutions in the resulting protein. These changes were not uniform along the gene but clustered in two of the exons—or coding sequences—exons 7 and 17. In several species of echolocating bats there was clear evidence for accelerated evolution. Leaf-nosed bats and all vesper or evening bats showed different mutations in exon 7; several species of echolocating bat shared amino-acid substitutions with whales—which also exhibit vocal learning; and in exon 17, amino-acid variation was considerable in all echolocating bats with up to eight different substitutions noted. The most divergent species was the slit-faced bat, *Nycteris*, which is commonly found throughout Malaysia and Indonesia. Interestingly, these bats emit very short multi-harmonic sonar calls and target their mainly insect prey, not by sonar bounce-back, but by eavesdropping on the tiny sounds they make. They conclude that echolocation in bats has involved the evolution of *FOXP2* and that downstream cascades of genes under its regulatory control are involved in the development and maintenance of this very complex form of sensorimotor coordination. Humans, song-birds, and bats have at least one thing in common not shared with chimpanzees: complex forms of communication requiring fast and dexterous motor coordination of the muscles associated with sound (or ultrasound) production from the larynx. Examining *FOXP2* variation more carefully across a broader range of species will help to decide whether its role is specific to vocal learning, articulation, or the complexity of the signals animals emit. This more circuitous route may lead us to a better understanding of the nature of language and why we humans can talk, whereas chimps cannot.

CHAPTER 3

Brain-builders

In 1967 the Pakistan government built the world's twelfth biggest dam across the Jhelum river, south of Islamabad, to create a vast reservoir and hydro-electric power generation project. The resulting 100 square miles of lake engulfed two major towns, Mirpur and Dadyal, and some 280 villages. Over 100,000 people were displaced from their homes. Many of them were given work permits to come to the United Kingdom, with the result, today, that most of the Pakistani communities in big, former mill towns, like Bradford in Yorkshire, are composed of extended families from the Mirpur area of Pakistan-controlled Kashmir.

Initially it was the men who came—to work in the mills. Then, gradually throughout the 1970s and 1980s, their families followed. There is a strong tradition of marriage within the clan, or biraderi, and marriage between first and second cousins is very common. Family members married related family members in these new British communities and, later, UK-born Pakistani men would bring in marriage partners from related families back in Pakistan. Occasionally there would even be double first marriages where the mother and father of one of two first cousins getting married were also first cousins. Over the years this has led to the creation of large numbers of super-families where the level of consanguinity is high. Consequently, the Pakistani community suffers from a medical load of genetic disease that is far greater than is experienced in the surrounding population. Dr Peter Corry, a paediatrician who runs the Child Development Centre at Bradford Royal Infirmary, estimates that some 140 recessive genetic disorders have been seen in the Bradford area over the last two decades, compared with an expected norm of about 20 to 30 in a typical health authority. (Remember gene mutations come in pairs—dominant and recessive. In traits or diseases that are due to the effects of a mutation in a single gene, if that gene is recessive you will only risk developing the disease if both your parents carry the recessive gene mutation, which will make you homozygous. The very high level of consanguinity among these Pakistani families has raised the chance that this will happen.)

Many of these disorders have led to neuro-degenerative conditions; there are high rates of cerebral palsy, Joubert's syndrome, and microcephaly, where children are born with very small heads and moderate to severe mental handicap. The typical weight of an adult male microcephalic brain is about 430 grammes—not unlike that of a chimpanzee—compared to 1,450 grammes for a normal human adult. Microcephalics have small heads because the cerebral cortex, that much-folded part of our brains that is responsible for all higher-order mental function, fails to develop properly.

In the late 1980s a clinical geneticist, Geoff Woods, took up a post at the St. James University Hospital in nearby Leeds. Bradford was on his patch. Microcephaly was a research priority but he couldn't make very much headway with the Pakistani community because they kept themselves and their problems very much to themselves, didn't trust the medical and care services that much, and were generally poor attenders at hospital and doctors' surgeries. Children born with neurological defects were simply absorbed by and cared for within the families—as they had always been. Together with Peter Corry and a second colleague, Bob Mueller, they began setting up a network of Asian counsellors and local Pakistani GPs to improve levels of contact and trust and gradually more and more families were located and began trooping in. By the mid 1990s, recalls Peter Corry, they had over 40 microcephalic children on their books.

Several of the families had two, even three, affected members. To Geoff Woods, microcephaly was clearly genetic and recessive and it became imperative to discover the genes responsible. Because the brains of microcephalics grow quite normally up to the third trimester of pregnancy no early ultrasound scan can detect the abnormality. What was needed was a genetic test that could detect problems in very young embryos and, more importantly, lead to a genetic screening test to help young Pakistanis avoid consanguineous marriages that were likely to lead to microcephalic children.

Woods and his colleagues started with one of the largest super-families, which had five affected members between 13 and 28 years old. They later added a further eight consanguineous families containing a total of twenty-three affected members. They took blood samples and began the gene hunt. Their initial hypothesis was that they would find one gene and they used standard genetic linkage analysis to discover it. Linkage analysis allows you to narrow down the search area from the entire genome, spread out over all the chromosomes, to roughly where the gene is situated. It depends on the use of a known library of genetic markers: known, recognizable DNA sequence variants, which are spaced out along the chromosomes at intervals, a bit like mile markers on a roadside. In a large family it is possible to map the occasions when a particular marker and the affected condition turn up together in the same individual because the gene for the condition is located close enough to the marker for them both to be carried into the next generation on the same chromosome, surviving any recombination, or crossing-over, that occurs between the chromosomes during sexual reproduction.

By comparing results, originally from five members of the first family, and then by pooling families 1 and 2, they discovered a common homozygous region equivalent in length to 13 million basepairs of DNA, on the short arm of chromosome 8. The gene they were after had to be inside this area. They named the locus MCPH 1 (MicroCePHaly 1).

They were very glad they had selected the largest families. In their first paper, in 1998, reporting the chromosome 8 locus, they mentioned they had found evidence for the presence of several other loci and that figure has now grown to eight. Genetically, microcephaly has proved to be very heterogeneous. Had they originally started looking in small, nuclear families it could well have been that each family had a different genetic root for the disease—the families would simply have confounded each other, leaving the scientists baffled.

Of course, identifying a locus for the disease is not the same as identifying the gene. If the region they had pin-pointed was equivalent in length to Oxford Street in central London, they were now looking for one kerb-stone somewhere along it. This was before the first detailed reads of the human genome had been published to aid their search—they had to go in with blunter tools. They took two of the families that showed linkage to MCPH 1 and reconstructed the locus in more detail, using an array of known and novel marker sequences. They overlapped the maps from parents and affected individuals from both families and looked for the most discrete area of overlap. It turned out to be a relatively tiny stretch of some 2 million nucleotides of DNA which contained only two known genes. One of them, angiopoietin-2, was screened for mutations but none were found. Attention shifted to the second gene, *AX087870*, about which nothing was known and they found a single mutation in one of the coding areas of the gene which prematurely stopped the gene coding for protein at that point—severely truncating the resulting protein. All seven affected individuals in the two families were homozygous for the mutation, and all the parents were heterozygous—and therefore carriers of the mutation. They had found it! They duly christened the protein 'microcephalin'.

From the very beginning they were alert to the possible evolutionary implications of microcephaly genes. As they point out, brain volume in microcephaly is roughly comparable to that in early hominids and it had already been proposed that microcephaly was an atavistic disorder—a reversion to primitive characteristics present in distant ancestors. Indeed, for this very reason, microcephaly has been at the heart of one of the most vituperative palaeoanthropological spats in recent memory—regarding the true nature of the 'Hobbit', *Homo floresiensis*, unearthed on Flores island in Indonesia in 2004. The Indonesian anthropologist Teuku Jacob has maintained that the creature is a dwarf *Homo sapiens* who suffered from microcephaly— his brain was scarcely bigger than an orange. The Australian scientists who had actually dug up the 'little man' in the Liang Bua Cave were incensed—they had deduced, from myriad measurements of the other skeletal remains in the cave, that the 'hobbit' was a distinct human species. Dean Falk, an anthropologist from Florida State University, compared endocasts—plaster-casts of the inside of skulls—from 10 known microcephalics with that of the 'hobbit' and pronounced the hobbit skull to be different. Meanwhile, a team of German neuroscientists and anthropologists, comparing the 'hobbit' with 27 microcephalic endocasts, placed the 'hobbit' in the middle of the range of human microcephalic skulls—they couldn't rule out the possibility that the 'hobbit' was a deformed, malnourished, human microcephalic. The controversy continues to rumble on, with Falk, to date, having the last word, in concluding that the 'hobbit', as the Australians originally claimed, is an extinct form of *Homo erectus*. The whole farrago may tell us more about palaeontologists than it does about human evolution, and it certainly shows up the extreme difficulty surrounding the interpretation of fossilized skeletal remains.

What role has the gene *Microcephalin* played in human evolution? Andrew Jackson and the team set out to pin down exactly what the gene does and where it is active. Experiments with human foetal tissue showed it was expressed in human brains, and work on mouse brains pinpointed its expression during neurogenesis, when neurons are being formed from precursor cells in the lateral ventricles of the fore-brain. These are the tissues from which neurons migrate to form the cerebral cortex. The number of cells the cortex ends up with depends upon how long this process of neurogenesis is, and there is a trend to longer periods of neurogenesis along a line from rodents to primates.

What about the other loci? Here Christopher Walsh, a neuroscientist from Harvard University, enters the picture. Walsh has always been tuned in to the possible evolutionary implications of his work:

> Most neuroscientists would sit around on dorm floors until 2 o-clock in the morning talking about 'Wow! Consciousness! What a wonderful thing our brain is … what makes us different to all the other animals?' A lot of the people who were having those bull-shit sessions late at night in college ended up in neuroscience, asking these evolutionary questions.

His lab had already started building up a list of brain-building genes that might have important evolutionary tales to tell and he had already started mapping microcephaly genes in Arabic families from Jordan. When he stumbled over a family that suggested linkage to one of the loci that Geoff Woods had already discovered Walsh decided to lure him into collaboration:

> We said, well Geoff already published this locus, we should work with him, he's got good reagents, and so we provided the family to him and worked together on the mapping.

Walsh paid for Woods to travel to the US and the deal was sealed in the stands at Fenway Park during a Boston Red Sox baseball game.

The locus concerned was on chromosome 1 and was the fifth out of eight microcephaly loci eventually found, MCPH5. It is by far the most prevalent cause of microcephaly in an arc from the Middle East, through Pakistan to India—all areas where first-cousin marriages are common. They narrowed the gene search down to an area of 600,000 nucleotides which recently published genome databases suggested contained four known genes. One of them, ASPM, contained a different mutation in each of the final four families from which the last tranche of data had been pooled. All the mutations were similar in that they introduced a premature full-stop into the reading frame of the gene—truncating the protein that resulted from that gene's transcription. ASPM is the human version of a gene that appears throughout the animal kingdom. In fruit flies it is named asp, which stands for 'abnormal spindle'—we'll see why in a moment. Across the animal kingdom, most of the areas, or domains, of the protein differ very little—they are remarkably conserved. Exceptions are the so-called IQ domains. IQ here is not named for Intelligence Quotient, but because I and Q stand for two of the amino-acids in the resulting protein chain—though, ironically, the number of repeats of this IQ domain increases from insect to mammal as brain size increases. The humble nematode, for instance, has only 2 IQ repeats, the fruit fly boasts 24, the mouse 61, and ASPM in most other mammals, including man, has a massive 74.

Work with mice and fruit flies has shown that the various species variants of ASPM are crucial for control of the cell division by which germ cells, or neuroepithelial cells, proliferate. This occurs through a process called mitosis. Mutations in asp stop this proliferation in its tracks, resulting in poor development of the brain and central nervous system, by interfering with the precise geometry with which mitosis occurs. A brief description of mitosis will help to explain just how critical the effect of this gene is.

At the beginning of mitosis, each one of the pair of each chromosome divides into two identical daughter chromatids. Then a structure called the mitotic spindle forms out of micro-tubules that taper out to two opposite poles which will become the site for the two nuclei of the daughter cells. The chromatids align at the centre of the mitotic spindle, to which they attach, and migrate out along it to its poles such that each of the pair of daughter chromatids migrates in the opposite direction to the other—the resulting two nuclei will therefore end up with exactly the same genetic complement as the parent. The ASPM gene family seem crucial to the plane of the mitotic spindle—they hold it in a horizontal position. Once the chromatid migration has occurred, a vertical cleavage furrow—perpendicular, therefore, to the plane of the mitotic spindle—develops to cut the parent cell neatly into two identical, symmetrical daughter cells. These remain in a totipotent condition which means they can divide repeatedly—the population of germ cells therefore doubles at each division. Any departure from this symmetry will result in only one of the two daughter cells remaining in the progenitor condition with the second cell losing its progenitor status and becoming a neuron, which then begins to migrate to take its place in the developing layers of the cortex. So any departure from symmetry curtails the active proliferating life of the neuroepithelium.

In a beautiful piece of recent research, a group of German scientists have shown just how crucial this business of symmetry is—and in doing so have pointed the way to why selection pressure on genes like ASPM is so intense. This ability of the germ cells to keep on dividing depends much more on the vertical symmetry of the cleavage furrow than was previously thought. This is because, at the outset of mitosis, these cells elongate about their vertical axis—it helps to imagine them growing from a squat shape into something like a row of beer bottles. At the top of each cell sits a tiny structure called the apical cap in which a ring of molecules surrounds a central blob of a protein called prominin 1. When the vertical cleavage furrow forms it runs right across the cell, apex to base, and cuts through the apical cap. If the mitotic spindle has been held exactly horizontal,

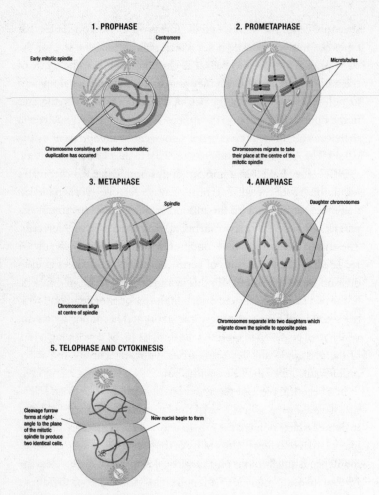

FIGURE 1. The stages of mitosis

the cleavage furrow will be exactly vertical and the apical cap will
be cut neatly in half. Each daughter progenitor cell will have half the
original complement of prominin 1 and this equal share-out ensures
that both daughter cells remain in the totipotent state. At each successive division the prominin 1 is halved again until there is no longer

a critical amount left to keep the resulting cells in their progenitor state—they start to form neurons. Also, should the cleavage furrow be off vertical by as little as a few degrees at any one generation of cell division, one cell will get less prominin 1 than the other and, again, will lose its progenitor status—it will develop into a neuron. In this way, prominin 1 is a clock for cell division, cortex size, and neuron number. This explains why *ASPM* mutations cause microcephaly and why beneficial mutations in *ASPM* could have been selected. Any expansion in cerebral cortex means, up to a point, keeping as many cells in the neuroepithelium in their totipotent condition for as long as possible—squeezing proliferation power out of the neuroepithelium. The longer the population of totipotent progenitor cells can be maintained, the more cortical neurons will eventually be produced, and the more cortical expansion will be supported. Theoretically, any mutation in *ASPM* which improved its control over the plane of the mitotic spindle should be evolutionary gold-dust. It could help to build bigger brains.

So, is *ASPM* one of the genes that made us human? Geoff Woods, Chris Walsh, and Nataly Kouprina, of the National Cancer Institute in Bethesda, Maryland, compared rates of evolution of the gene among rhesus macaques, orang-utans, gorillas, chimpanzees, and humans, to back their hunch that DNA sequence evolution somewhere in the IQ repeats of the gene would prove important. They used a classical method to test for positive selection: the ratio of non-synonymous to synonymous mutations in each of the domains of the gene. Not every point mutation—single substitution of a DNA nucleotide—results in a change in the corresponding amino-acid of the protein which is produced by that gene. If this is the case, the mutation is silent—it is almost as if nothing had occurred. Since no change is made to the structure of the resulting protein its function remains the same—the change is neutral. The mutation is therefore said to be synonymous. However, in a non-synonymous mutation, a change or substitution of one nucleotide in the genetic sequence of the coding part of a gene results in a change or substitution of an amino-acid in the molecular chain of the resulting protein. This may well change the biological function of that protein which may be either advantageous or deleterious for the organism—it will be selected for or against. A ratio of non-synonymous to synonymous mutations (Ka/Ks ratio for short) that is greater than 1.0 is usually taken as evidence that positive selection has occurred.

The team found that most parts of the ASPM protein had been highly conserved throughout primate evolution—there were very few changes of any note. The rapidly evolving sites were indeed mainly concentrated in the IQ repeats. Remember, the number of IQ repeats is the same throughout the primates—it was the pattern of non-synonymous, or positively selected, mutation within the IQ repeats that was different. They noted particularly strong signatures of selection in the IQ domain in both gorillas and humans, with a lesser signal in chimpanzee. However, looking deeper into the primate family tree revealed that evolution of ASPM had begun to accelerate some 7–8 million years ago in that part of the primate family tree after the split from Old World monkeys and orang-utans. This includes gorillas, chimps, and humans, and their common ancestors. This was well before the runaway period of human cortical expansion which began only 2.5 million years ago and ended only a quarter of a million years ago. The strong difference in rates of ASPM evolution between orang-utans on the one hand and gorillas/chimps/humans on the other is not reflected in the sizes of primate brains today—where male orang-utan brains weigh in at 430 grammes, gorillas 530, and chimps 400. They concluded that their results best fitted the idea that ASPM has had some role in changes in brain anatomy that are more subtle than simple expansion and that most of these effects occurred very early in hominoid history. To this day no one has come up with an answer as to how increasing the number of IQ domains, or making changes within IQ domains, relates to cortical size.

The arrival of two more American research groups on the scene caused things to get much more exciting, very quickly. Jianzhi Zhang, from the University of Michigan, began his own investigation by sequencing and comparing the *ASPM* gene in a range of human ethnic groups to the chimpanzee and orang-utan. Again, he compared the ratio of non-synonymous to synonymous mutations and found that the ratio was lowest in the orang-utan branch, at 0.43, higher in chimpanzee at 0.66, and highest in the human branch at 1.03.

At any nucleotide in the DNA sequence of a gene there may exist in the population a number of variants. These are called polymorphisms—in this case single nucleotide polymorphisms (SNPs) called 'snips' for short. Some chromosomes in the population will carry the mutated and therefore substituted nucleotide at this point, others the ancestral nucleotide. If selection has strongly favoured the altered protein that results from a non-synonymous mutation, that polymorphism will increase in prevalence within the population right up to 100%. At this point it is said that the mutation has become fixed, because all individuals carry it. A bit of mathematical analysis allowed Zhang to posit that, of the 16 non-synonymous mutations unique to the human lineage, there was overwhelming proof that 12 of them had risen to fixation due to the effects of positive selection. This period of selection, for Zhang, had to be between 6 million and 100,000 years ago—with the nearer date coinciding with the time it is thought that *Homo sapiens* migrated out of Africa. Zhang's conclusion, that there had been substantial selection on *ASPM* peculiar to the human line after the split from the common ancestor with chimpanzees, was much stronger than the more cautious conclusions of Geoff Woods, Chris Walsh, and their team. The Woods group grumbled that, since Zhang had not included gorillas in his comparison, he had missed the earlier burst of *ASPM* evolution during the period before hominid brain expansion began. But Zhang's mutations had caused amino-acid substitutions and therefore advantageous protein structural changes that were not related to the IQ domains that Woods had concentrated on. They had occurred elsewhere on the ASPM protein. Something else seemed to be going on.

Enter the third research group, of Bruce Lahn and his colleagues, from the Howard Hughes Medical Institute of the University of Chicago. Taking exactly the same approach, using the ratio between non-synonymous and synonymous mutations and comparing species of primates from owl monkeys, through macaques, gibbons, and up to orang-utans, gorillas, chimpanzees, and man, they found that there was a trend to higher Ka/Ks ratios the nearer one got to man. The great ape lineages had higher Ka/Ks ratios than the lower primates and there had been two big spurts of positive selection, when the Ka/Ks ratio had greatly exceeded 1.0, in the lineage leading to the four great apes, and even greater still, in the most recent lineage, that leading to humans from the common ancestor with chimpanzees. Here was evidence for dramatic and sustained positive selection on ASPM throughout primate evolution, and especially over the last 6 million years.

When they ran a sliding window of Ka/Ks ratios along the whole length of the ASPM gene they found peaks and troughs corresponding to highly conserved and highly variable areas. The two areas that showed the strongest evidence for positive selection were those that corresponded to the IQ domains and other areas of the protein that bind to the mitotic spindle. This corroborated both Zhang's and Wood's findings that evolution *within* IQ domains, but not *of* IQ domain number had been important. The clock of ASPM evolution, according to Lahn, had been ticking furiously since the split with the common ancestor. There was clear evidence that 15 mutations had been driven to fixation over the last 5+ million years—corresponding to one adaptive amino-acid change in ASPM for every 300,000 years since we diverged from chimpanzees. A furious rate.

The same group also turned their attention to *Microcephalin* and reported that it had accumulated 45 advantageous amino-acid changes during the 25–30 million years in the line from the early simian primates (Old World monkeys like colobus) through the great apes to humans. That is one substitution for every 600,000 years—again, quite a pace for Mother Nature. This trend was far more pronounced than that for any other mammals. By far the greatest selection spurt for *Microcephalin* occurred relatively early on in the evolution of the higher primates, in the lineage leading to the common ancestor of all the great apes, suggesting an ancient and modern story with respect to *Microcephalin* and *ASPM*, whereby *Microcephalin* was more implicated in brain size increase at the origin of the great apes and *ASPM* with human encephalization over the last few million years.

Thus far the story on microcephaly genes and their role in human brain evolution is uncontroversial. However, in 2005, Lahn's group produced two startling claims, one for *Microcephalin* and one for *ASPM*, which not only suggested that human brains are still evolving—and that these two genes are at the heart of that evolution—but immersed Bruce Lahn in red-hot scientific controversy amid accusations that he was conducting racist science. The controversy began with two papers published back-to-back in *Science* journal in September 2005.

In the first, lead-authored by Patrick Evans, the team reported that they had identified a genetic variant of *Microcephalin* that had arisen in modern human populations as recently as 37,000 years ago. This is the period in prehistory when modern humans were spreading rapidly throughout Europe, supplanting Neanderthals, and leaving traces of complex culture like cave paintings and numerous artefacts. The team maintained that the gene had increased its frequency in the human gene pool far too rapidly for it to be explained by neutral selection—what geneticists call random drift. It had been positively selected.

The approximate age of this variant, at 37,000 years, substantially post-dated the emergence of *Homo sapiens* out of Africa and the new variant did not appear to have uniform distribution across all modern ethnic groups. Using a wider set of 1,184 individuals from all over the world they found that sub-Saharan populations had much lower frequencies of it than Asian or European populations, where the frequency had risen to a massive 70%.

Lahn's group clearly suspected that the human brain had been the target of selection for either brain size or some aspect of cognition. They hypothesized that this *Microcephalin* variant might have changed the pattern of cortical growth and development in such a way that it might be possible to measure differences in brain size and cognition between human ethnic groups today.

The companion paper on *ASPM* was even more dramatic. The same research team, this time led by Nitzan Mekel-Bobrov, claimed to have found a variant of *ASPM* that had arisen a mere 5,800 years ago and had since swept to high frequency under strong positive selection. This suggested, they said, that the human brain was still evolving. They sequenced the gene, this time in 90 ethnically diverse individuals, and found one variant that had a relatively high frequency of 21%, compared to over 100 other variants all with frequencies less than 3%. This variant contained two non-synonymous mutations in the same area where earlier work had shown strong selection pressure in the lineage leading to humans. In this case the frequency of the variant was much higher—at over 40%—in Europeans and Middle Easterners (including Iberians, Basques, Russians, North Africans, and South Asians) compared with sub-Saharan Africans. They noted that the approximate date of 5,800 years ago fitted quite well with the 'emergence and spread of domestication from the Middle East at around 10,000 years ago, and the rapid increase in population associated with the development of cities and written language 5,000 to 6,000 years ago', though they added in final conclusion, 'the significance of this correlation is not yet clear.'

The two *Science* papers created a press furore. Among the welter of articles, two of America's most prominent right-wing writers and bloggers—Steve Sailer, a regular contributor to *The American Conservative*, and John Derbyshire, who writes for the *National Review*—bent Bruce Lahn's discoveries to their own political agendas. For them, the significance was obvious. Sailer had just written a controversial article on the debacle of New Orleans in the aftermath of hurricane Katrina. The plight of the thousands of overwhelmingly poor black survivors he put down to a lethal cocktail of cavalier indifference from the Bush regime in Washington, and a perceived essential weakness in the black population itself:

> It also should have been expected that a large fraction of New Orleans's lower class blacks would not evacuate before a disaster. Many are too poor to own a car, or too untrustworthy to get a ride with neighbors, or too shortsighted to worry…Judging from their economic and educational statistics, New Orleans' blacks are not even an above-average group of African-Americans, such as you find in Atlanta or Seattle, but more like Miami's or Milwaukee's. About half are below the poverty line. With the national black average IQ around 85, New Orleans' mean black IQ would probably be in the lower 80s or upper 70s

This article preceded the duo of Lahn papers in *Science* by five days and linked what had happened in New Orleans to the highly controversial research on race, IQ, and criminality by maverick psychologist J. Philippe Rushton, from the University of Western Ontario in Canada. Rushton has repeatedly put forward evidence that IQ, among the major ethnic divisions in the world, ranks Asians > Europeans > sub-Saharan Africans. Lahn's two papers had reflected this IQ ranking in the distinction between ethnic groups in the frequencies of the new variants of *Microcephalin* and *ASPM*. On 11 September, Sailer justified his remarks by referring to Lahn's papers:

My 'New Orleans Nightmare' article was exposed to a lot of scientifically illiterate loathing. But I had a couple of genetic aces up my sleeve: I knew that later in the week, the leading American scientific journal, Science, would publish two blockbuster papers by U. of Chicago geneticist Bruce T. Lahn ... These findings are far from definitive on the IQ innateness question. But I can assure you there will be more announcements to come about these two brain genes that will be most interesting. And more brain genes with politically incorrect racial distributions are likely to follow

A little later, in November, the columnist John Derbyshire spelt out the impending unpopularity of Lahn's claims because of their apparent conflict with deeply felt constitutional ideas in America about the fundamental equality of all men:

BOMBSHELL PAPERS

Two papers published in the Sept. 9, 2005, issue of Science illustrate my point ... Mysteriously, both these variants seem to be scarce in sub-Saharan Africa. But ... but ... but ... haven't talking heads on TV science programs and in the pages of respectable newspapers been telling us for years that there are absolutely no significant genetic differences whatever between human groups defined by common ancestry? Yes; but you see, they have been lying through their teeth. To be fair, those talking heads have been lying with the best of intentions. Ours is a nation founded on the proposition that all men are created equal. Of course, nobody ever supposed that to mean that we are all equally tall, equally strong, or equally clever; but *if different human groups, of different common ancestry, have different frequencies of genes influencing things like, for goodness' sake, brain development, then our cherished national dream of a well-mixed and harmonious meritocracy with all groups equally represented in all niches, at all levels, may be unattainable.*

The University of Chicago joined in the enthusiasm for Lahn's brain evolution genes by applying to patent a test to identify carriers of the new variants. However, rumblings soon began from the scientific establishment that, in even speculating that there might be links between the arrival of these new variants in the human gene pool and important stages in recent human cultural evolution, Lahn had wildly overstepped his data. Michael Balter, in *Science* magazine, reported David Altshuler, of the Broad Institute in Cambridge, Massachusetts, as saying that the links to cognition were wild speculation and that scientists had a responsibility to think about how society would interpret such work. One could easily have anticipated controversy and 'there was no evidence whatsoever that these genetic variants have any effect'. Balter also reported veteran population geneticist Richard Lewontin as saying 'These two papers are particularly egregious examples of going well beyond the data to try to make a splash', and archaeologist Scott MacEachern saying the archaeological links in the paper were simplistic and outdated, while one of Lahn's collaborators on both papers, Sarah Tishkoff, who had previously said that the signature for positive selection on *ASPM* was among the highest she had ever seen, was reported as 'distancing herself from the project'.

Antonio Regalado, in the *Wall Street Journal* in June 2006, conducted his own investigation into the backlash and turned up strong criticism from geneticist David Goldstein of Duke University and Mark Stoneking from Leipzig—both of whom found the links to brain and behaviour unconvincing. According to Regalado, Pilar Ossorio, professor of medical ethics at the University of Wisconsin, drew explicit links to Charles Murray's and Richard Herrnstein's enormously controversial book *The Bell Curve*, published in the mid 1990s, which argued that lower performance on IQ tests among African-Americans had a genetic component. The director of the University of Chicago's patent office, Alan Thomas, told Regalado that he was dropping the patent application on Lahn's work on a DNA-based intelligence test, saying 'We really don't want to end up on the front page...for doing eugenics.'

Perhaps the most important criticism came from Francis Collins, one of America's top genome scientists and director of the genome programme at the National Institutes of Health. He is quoted as saying, 'This is not the place you want to report a weak association that might or might not stand up.' According to Regalado, Collins obtained advance copies of the two *Science* papers and circulated them to top population geneticists for criticism.

That criticism—in science print—was not long in coming and centred around Lahn's speculation that the new variants of *Microcephalin* and *ASPM* should show up as measurable differences in human populations today in either brain size or cognitive ability. Here was bad news for Bruce Lahn. In May 2006, a group headed by Roger Woods, of the University of California, Los Angeles, genotyped 120 individuals of different ethnicity and measured their brain volumes using MRI scanners. They found absolutely no association between the variant alleles and increases or decreases in brain volume and concluded that 'the selective pressure on these genes may be related to subtle neurobiological effects or to their expression outside the brain'.

Next, it was Philippe Rushton's turn to report in. Remember, Rushton has been notorious for years for his persistence with research into racial differences in intelligence, criminality, and human male reproductive strategies. He has survived repeated demonstrations in Canada and calls for the withdrawal of his tenure at the University of Western Ontario. He looked to see if there was a correlation between the distribution of the recent gene variants and head circumference, performance on a battery of IQ tests, and questionnaire tests of social intelligence. He tested 644 Canadian adults: most were white, with smaller numbers of orientals and blacks. Although confirming Lahn's results with regard to the different frequencies of the two variants among human populations he could find no relationship between those genes and any of his measures.

In January 2007, Lahn finally threw in the towel. A large international combined effort, lead-authored by Nitzan Mekel-Bobrov, reported on a large-scale association study on over 2,000 human subjects, between the *Microcephalin* and *ASPM* variants and several measures of IQ, and found absolutely nothing. Lahn's ambitious, quick-and-dirty route to brain evolution had finally run into the sand.

The idea that recent evolution of *Microcephalin* and *ASPM* could be reflected in some gross parameter like brain size or IQ has been fatally undermined. Indeed, Geoff Woods—the British geneticist who did the original work on microcephalic families—has always found it an unlikely assumption. He thinks that it might always prove impossible to find the real functional differences these gene variants cause. Given the very conservative nature of brains it is highly likely that those differences would be very small. 'Suppose one mutation caused a 1% difference in the kinetics of some kind of neurochemical reaction in the brain,' he says, 'you would never find it.'

We had to wait until April 2007 for the most trenchant criticism to date of Bruce Lahn's claims for one of the two microcephaly gene variants. It came from a consortium of genome experts, mainly based in Cambridge, Massachusetts, lead-authored by Fuli Yu. They concentrated their fire on *ASPM* and not only questioned Lahn's dates for when the variant had arisen, at approximately 5,800 years ago, but disputed that it had ever encountered any exceptional selection pressure. Their arguments, if correct, would blow Lahn's evolutionary scenario wide apart.

Yu pointed out that Lahn's group had identified the variant allele of *ASPM* and estimated its frequency at greater than 40%. They had then used computer simulations of various demographic histories and shown that none of them could have produced a variant of such a high frequency. But would *ASPM* prove to be really exceptional when compared, empirically, with real genomic data? When Lahn's group had published their results in *Science* the main International HapMap library of human genetic variation had yet to be fully published. This, in effect, scanned the genomes of some 270 individuals spread across four main population groups in the world—in Africa, Europe, Japan, and China. The HapMap consortium selected one single nucleotide polymorphism (snip) from every 5,000 nucleotide bases across the genome, and by August 2006, they had accumulated more than 10 million snips of which more than 40% appeared to be polymorphic—different alleles existed among human populations for each snip. Yu compared ASPM with this publicly available snip data and the signature for selection for Lahn's ASPM variant simply didn't stand out from the crowd. Furthermore, Lahn had stated that the allele frequency for the ASPM variant differed greatly between his sub-Saharan African group, and the European and Asian groups (the foundation for strident claims of 'race science'), but Yu and his team found that over a third of the snips they looked at had at least as strong a differential signature for selection. Finally, they disputed Lahn's estimation of the age of his ASPM variant, calculating that it had arisen in European history 'at least tens of thousands of years ago and possibly more than 100,000 years ago'. They then rubbed Lahn's nose in it by pointing out that the ASPM variant occurs at 50% frequency in Papua New Guinea Highlanders who are thought to have diverged from Europeans over 40,000 years ago. So much for thoroughly modern ASPM-mediated brain evolution!

Lahn was unimpressed and, together with Nitzan Mekel-Bobrov, mustered a typically robust response. At the heart of it was strident criticism of the data Yu had relied upon which Lahn claimed contained an innate bias that inflated estimates of positive selection among the snips in the HapMap collection. When that was corrected, ASPM was proved to be exceptional after all. Lahn backed this up by comparing his data for the ASPM variant to data on 289 genes from 47 ethnically diverse individuals in the so-called Seattle Project. Many of these snips are thought likely to have been under positive selection because they occur in genes involved in the immune system, an evolutionary hot-spot. Sure enough, only 1.7% of this highly skewed sample of genes (that's only 5 out of 289) showed evidence for evolution that was equal to or greater than that for *ASPM*.

As for New Guinea—the island had hardly been in splendid isolation during human prehistory. It had endured successive waves of immigration since 40,000 years ago, including one significant migration only 3,500 years ago—there was no contradiction at all. Yu had relied on a biased data-set and poor demographics, Lahn's data was robust, and the evidence for strong recent selection on *ASPM* still stood, they concluded.

The jury is still out on this rather extraordinary controversy. According to Chris Walsh, who was a co-author with Yu, the same group has taken a preliminary look at Lahn's evidence for very recent evolution of *Microcephalin* 'and the evidence so far doesn't look too good for Bruce'. However, if it is based upon the same allegedly flawed approach nothing is more certain than that this determined, brilliant researcher will bat it back with the same ferocity. The main casualty in the affair is that, horrified by the politically charged quagmire he found himself in, Bruce Lahn has recoiled from the quest to find 'the genes that made us human' and is diverting into the relatively calmer waters of stem cell research.

While his quick-and-dirty route from gene evolution to how our minds work didn't pan out, two researchers from the University of Edinburgh have given Bruce Lahn's cloud a silver lining by discovering an extraordinary link between his two recent variants of *Microcephalin* and *ASPM*, and language. Bob Ladd, a linguist, was joined by genetics research student Dan Dediu, to examine the world distribution of the new variants of *ASPM* and *Microcephalin* discovered by Bruce Lahn. There was a superficial resemblance between the world distribution of older alleles of *Microcephalin* and *ASPM* and tonal languages, and Lahn's new variants and non-tonal languages. For instance, tone languages are the norm in sub-Saharan Africa and are very common in continental and insular south-east Asia, relatively common in central America, the Caribbean, and the Amazon Basin, but are rare in the rest of Eurasia, north Africa, and Australia. Wherever tone languages are common the new gene variants are rare.

As Ladd explains, the world is roughly divided between non-tonal languages, like English, which mean what they say, and tonal languages, like Chinese, in which meaning depends on how you say it. All languages, he says, use consonants and vowels to distinguish one word or grammatical category from another but, in addition, so-called tone languages, like Chinese, use pitch for this purpose as well, whereas non-tone languages like English use pitch only at the sentence level to convey emphasis or emotion. In tone languages, pitch is organized into tone phonemes that are functionally comparable with consonant and vowel phonemes. In Chinese, for example, the word 'huar' spoken with a high pitch means 'flower' and with a falling pitch, means 'picture'; and in the African language, Yoruba, the word 'igba' can have four meanings, dependent upon tone, such that, when pronounced with a low to high intonation it means a species of tree, when pronounced with an even mid-tone it means an amount, '200', when pronounced with a mid-rising-to-high intonation it means 'gourd', and when pronounced sonorously low it means 'time'.

Although there will never be prominent 'genes for language', or, more specifically, 'genes for Chinese', there is every reason to suspect that genes do make small individual contributions to language in what are almost certainly multi-genic systems. Certain genes may very well, Dediu and Ladd argue, underpin aspects of language such as language development delays, aptitude for second language, discrimination between foreign speech sounds, recognition of words in noise, and other factors.

Their argument is bolstered by one piece of work by Patrick Wong and colleagues from Northwestern University, who noted that some adults brought up on English found it more difficult than others to learn an artificial language that used pitch and tone, and that this group showed subtle differences in brain structure. Specifically, they had found that ability to learn a pitch-rich second language was related to the size of the left side of a bilateral brain structure called Heschl's Gyrus, which is part of the auditory cortex. This surprised them because, although previous work had shown Heschl's Gyrus to be important in processing sound—whether pitch is rising or falling, which direction it is coming from, and how loud it is, etc.—it had never been shown to be associated with speech. But this association proved to be so strong that the size of the left Heschl's Gyrus, as measured in a brain scanner, was fully predictive of performance on the artificial tonal language. This could be due to genetic effects and the idea is supported by the fact that musical pitch processing is partially heritable, as is absolute pitch.

To correlate the alleles of *ASPM* and *Microcephalin* with tonal or non-tonal language, Ladd and Dediu drew up a list of 49 of the human populations identified in both the Lahn papers. For each of these they listed the frequency of the two alleles they were interested in and a set of over 1,000 reference alleles from two huge world genetics databases. They then whittled down 141 linguistic features from the *World Atlas of Language Structures* to 24. These included:

[are there > 26 consonants?]
[are there velar nasals?]
[are there uvular consonants?]
[are there > 6 vowels?]
[does the language have a tonal system?]

and the values for all these characteristics were estimated for all 49 populations.

They then calculated the correlation coefficients between all alleles and all 24 language features. As they suspected, for the vast majority of genes and linguistic features there was virtually no correlation at all: the resulting bell curve of their data shows a steep peak at the centre—where the correlation is zero. But right out at the tail of the distribution, among a handful of genes, lie *ASPM* and *Microcephalin*, with very high correlations indeed.

But what does it all mean? Correlations by themselves, no matter how strong, do not prove a causal relationship. Apparent correlations between genetic and linguistic 'diversities' are usually spurious because they are due to underlying geographical or historical factors. But when Ladd took these out of the analysis the correlations survived, suggesting a real association in which the effects of these two genes on brain growth and development might affect language acquisition and processing. It may be, say Ladd and Dediu, that the relationship is purely incidental, though they doubt it. And it certainly doesn't mean that non-tonal language per se was the selection pressure for Lahn's variants. Neither does it mean, in any way, that tonal languages are somehow inferior to non-tonal languages—or that the populations that speak them are in some way inferior.

Dediu and Ladd propose that the new variants of *ASPM* and *Microcephalin* conferred some selective advantage on the human brain and that one of the by-products of this was some slight cognitive bias that turned out to make non-tonal languages slightly easier to hear, learn, or employ. The idea is one of gene–culture co-evolution, where a very small genetic/cognitive bias gives rise to a slight bias in use of language, which gains ground culturally such that it becomes normalized in some populations. Geoff Woods, in criticizing Lahn's hypothesized link between genetic variations in *Microcephalin* and *ASPM* and really gross effects like brain size, suggested the change in the brain that might have resulted from either of these mutations would be too small to spot. Dediu and Ladd seem to have found that needle in a haystack.

Because of the very fortuitous way that these brain-building genes have been discovered—as by-products of neurological research—it is likely that they will only feature in a minor way in the search for the genes that made us human. Today's genomics research uses ultra-modern technologies of statistical search and analysis and complex arrays of thousands of genes on a 'chip' to allow the comparison of vast tracts of coding and non-coding DNA, and is able to look for the tell-tale signatures of selection across whole genomes. It is like the difference between a lucky prospector spotting a glistering nugget of gold in the bed of a mountain stream and panning for gold on an industrial scale. These ultra-modern techniques, however, are doing much more than revealing a growing list of 'genes that made us human': they are showing that the picture is far more complicated and exciting than the role of mutations within genes alone. The genome is a jungle filled with exotic beasts and genome explorers are adding to the menagerie every year. Chimpanzee and human genomes will never seem the same again!

CHAPTER 4

The Riddle of the 1.6%

What is indisputable—far too many independent scientific studies have verified it—is that over very appreciable lengths of their respective genomes, humans and chimpanzees are very similar indeed. This is where the oft-quoted '1.6% that makes us human' comes from. Despite 12 million years of evolutionary separation, 6 million for each species since the split from the common ancestor, we are surprisingly similar in our genes. Yet we are palpably very different animals when it comes to our bodies and brains. So, how many genes *does* it take to make us human? A handful of pivotal mutations such as *FOXP2* and *ASPM*, or many more?

The conundrum was well expressed by Ann Gibbons, veteran writer for the journal *Science*, in an article in 1998 titled 'Which of Our Genes Make Us Human?' when she said:

> We humans like to think of ourselves as special, set apart from the rest of the animal kingdom by our ability to talk, write, build complex structures, and make moral distinctions. Yet a quarter of a century of genetic studies has consistently found that for any given region of the genome, humans and chimpanzees share at least 98.5% of their DNA. This means that a

very small portion of human DNA is responsible for the traits
that make us human, and that a handful of genes somehow
confer everything from an upright gait to the ability to recite
poetry and compose music.

Let us take Ann Gibbon's figure of approximately 1.5% DNA difference and run with it for a moment. If we assume there are around
25,000 human genes that 1.5% would translate into about 375 genes
that were different. But, according to geneticist Jun Gojobori the
figure is higher by a factor of ten. We know, from the consortium
of scientists who reported on the chimp genome in 2005, that of
the 3 billion DNA nucleotide base-pairs in both chimp and human
genomes 35 million have been substituted by mutation. But Gojobori calculates that only 50,000 of these substitutions are relevant to
evolution because they have occurred in genes and have resulted in
changes to the amino-acid sequence of the proteins those genes code
for. Furthermore, he estimates, only between 10.4% and 12.8% of the
total number of genes containing these so-called non-synonymous
changes survive and get selected for. That still leaves us with about
5,000 genes to supply the recipe to make a human out of something
that resembled a chimp.

Bruce Lahn has always believed that the human brain is an exceptional organ compared to the brains of other primates, and that, for
it to get so big, and to develop such exceptional cognitive abilities in
so short a time in evolutionary terms—a paltry few million years—
required sustained and intense selection on a large number of genes.
To test this assumption, Lahn, together with Eric Vallender, Steve
Dorus, and others, drew up a list, culled from gene databases, of 214
genes known to be involved in building and running the brain and
central nervous system. They compared all these genes between rats
and mice, which are separated by approximately 20 million years
of evolutionary time, but where brain size, as a proportion of body
size, has not changed at all, and humans and macaques, separated by
a similar amount of evolutionary time but where the relative brain

size of humans has dramatically increased—up to nine times. They showed that the proteins derived from these nervous system genes had changed amino-acid sequence considerably faster in primates than rodents, and, within primates, much faster in the lineage leading to humans than lineages leading from macaques to the rest of the non-human primates. In other words, not only have brain and nervous system genes evolved rapidly in primates, but they have disproportionately accelerated in the human lineage.

They went on to identify 24 high-flyer genes—the majority of which were involved in brain growth or behaviour. These included *ASPM* and *Microcephalin*. When they compared this suite of 24 genes in humans and chimpanzees they discovered that sequence divergence was significantly, sometimes considerably, higher in humans. So evolution rate is especially elevated in the lineage leading to humans and particularly in genes governing brain development.

In 2004, a group of Cornell University scientists led by Andrew Clark and Carlos Bustamente managed to align for comparison some 7,645 genes common to human, chimp, and mouse. They were looking for signatures of a history of positive selection as reflected in DNA base substitutions over time—genes that had been altered more than would be expected by chance during the 6 million years since chimpanzee and human ancestors diverged. It was the first genome-wide comparison of humans and chimps and showed that hundreds of genes showed a pattern of sequence change consistent with adaptive evolution. Genes involved in smell, digestion, hearing, long bone growth, and hairiness were prominent. Hearing was interesting in that a gene, *alpha tectorin*, which codes for a protein in the tectorial membrane of the inner ear, showed a particularly strong selection signature. Mutations in the human version of this gene cause a form of congenital deafness because they lead to a poor frequency response—making it difficult to understand speech. Although there are no good audiology studies of chimps, Clark and his team speculate that chimps may not be up to the spoken language comprehension test because of this hearing difference.

In an obscure but thought-provoking paper in 2005, a team of researchers based at Pennsylvania State University went about chimp–human genome divergence a completely different way by looking at the cart instead of the horse. They went to a data-bank and compiled a list of 127 proteins common to both human and chimp, representing a total of 44,000 amino-acids. The great play that has been made over the minute DNA sequence difference between humans and chimps, they reminded us, is misleading because those measurements have included the vast proportion of the human genome—98%—that is made up of non-coding DNA which can have little or no effect on the phenotype—the brain, body, or behaviour. But proteins do. They found that in only 20% of those proteins was the amino-acid sequence identical. Think about it—this means that 80% of the proteins were different in humans compared to chimpanzees! They haven't a clue, as yet, as to how many—or how few—of these altered proteins have changed their biological function in a way that has made chimps and humans different, and consider that even this 80% protein difference may not be enough to explain all the phenotypic differences obvious to them between humans and chimps. Perhaps, after all, they conclude, chimp–human differences boil down to a small proportion of crucial genes, or master genes; or something else besides evolution in the coding sequence of genes and the amino-acid sequence of proteins, must be going on.

It is clear that different methods of working out chimp–human differences at the level of genes or proteins are capable of yielding different results. It is proving a real headache figuring out which are the gene or protein differences that count. Yet, somehow, an apparent 1.6%, plus or minus, at the DNA nucleotide sequence level translates into either a human or a chimpanzee. If you believe that humans and chimps are extremely similar in their brains and behaviour then it will make sense that they are also very similar in their

DNA sequence. However, if you believe, as I do, that this cognitive similarity is often grossly exaggerated, then, I think you will agree, we have a problem. While not even the most trenchant 'chimps are us' lobbyist would suggest that the finer points of human culture are within range of chimps, what is frequently said is that you can discern the foundations for human empathy, social intelligence, technology, mathematical skills, and moral rectitude in chimpanzees. But to call the difference quantitative between alarm calls, food-specific grunts, whoops, and Shakespeare; between night nests and twig tools, and the A380 passenger jet; and between retribution and food-sharing, and Aristotle and Mills is, to my mind, stretching a point, and a bit of an insult to human ingenuity and culture.

How can we reconcile such yawning intellectual and cultural differences between humans and chimps with that paltry difference at the level of DNA nucleotides? Certainly, Ann Gibbons' 'handful of genes' does not seem enough. Perhaps the classical, simple, view that evolution occurs via mutations which change a letter or more of the DNA coding alphabet, causing a change in the amino-acid sequence of the resulting protein, in turn causing a biological change in the host organism which undergoes natural selection, is also an insufficient explanation, regardless of how many genes you invoke? How can two such closely-related animals, in genetic terms, unravel, so to speak, into such different creatures as measured by what they say, do, make, write, and legislate?

Very much the same thought occurred to two molecular biologists at the University of Berkeley in 1975—Allan Wilson and Mary-Claire King. The paper they wrote on the subject in *Science* in 1975 gets my vote for one of the most important scientific papers of the twentieth century. They did more than pose the above question: they showed us a very promising potential route out of the whole conundrum—a way of squaring the circle. However, to a very large extent, Wilson's and King's solution has lain fallow for 25 years. It is only recently, armed with ultra-modern genomic technology, that a number of contemporary scientists have been able to examine the genome in sufficient detail to decide the extent to which they were right. These results make clear just how prescient Wilson and King were and are getting us very close to understanding how such apparent—and the stress is on the word 'apparent'—genetic proximity can give rise to such manifold differences between the two species' phenotypes.

In 1975, of course, King and Wilson did not have the modern genomic technologies now available to us, allowing us to provide high-density reads of whole genomes. Their evidence for apparent proximity shows exactly how limited the technology then available to them was. They pointed out that, owing to the limitations of sequencing methods, it was not then possible to sequence the amino-acids of the larger proteins but that micro-complement fixation—a measure of the immunological distance between two proteins—indicated that while some proteins common to both species differed very slightly, others were identical. They calculated the average difference between chimp and human, for all the proteins for which these data existed, to be 7.2 amino-acid substitutions per 1000 sites, making the sequences of these proteins on average 99% identical.

They then compared the proteins produced by 44 genes common to chimps and humans by a method called electrophoresis which separates protein molecules out by how far they travel through a chemical gel in an electric field. About half the proteins were electrophoretically identical, the other half were not. Did the protein sequencing and the electrophoresis agree? From electrophoresis they calculated the degree of amino-acid difference per 1000 sites to be 8.2—very close. Both estimates, they said, indicated that the average human protein is more than 99% identical in amino-acid sequence to its chimpanzee equivalent.

They then compared nucleic acids by hybridization. Here the double strands of the DNA from both chimp and human are made to separate by heating them. When they cool down they will re-form double helices wherever sequence identity is high and matches can be made. Where chimp DNA fuses with human DNA, hybrid DNA is formed. The degree of hybrid match is estimated by its thermal stability compared to the separate DNA of each species. They concluded that:

> The genetic distance between two species measured by DNA hybridization and electrophoretic data indicate that chimps are as similar as sibling species of other organisms. Immunological and amino-acid sequence comparisons of proteins lead to the same conclusion. Antigenic differences among the serum protein of congeneric [members of the same genus] squirrel species are several times greater than those between human and chimpanzee, those between congeneric frog species are 20 to 30 times greater. The genetic distance between humans and chimpanzees is well within the range found for sibling species of other organisms. This intriguing result, documented in this article, is that all the biochemical methods agree in showing that the genetic difference between humans and chimpanzee is probably too small to account for their substantial organismal [living body] differences … So it appears that methods of evaluating the chimpanzee–human difference yield quite different conclusions.

This is exactly the same conclusion reached by Maurice Goodman in the 1980s and 1990s using similar biochemical technology. Goodman's solution, you will remember, was to junk conventional taxonomic criteria in favour of a new molecular taxonomy which allowed him to re-draw the family tree of the apes so as to put chimps and humans in the same genus—*Homo*. But King and Wilson continued to puzzle over the discrepancy between molecular and morphological descriptions of the two species:

> The molecular similarity between chimpanzees and humans is extraordinary because they differ far more than sibling species in anatomy and way of life. Although humans and chimpanzees are rather similar in the structure of the thorax and arms, they differ substantially not only in brain size but also in

the anatomy of the pelvis, foot, and jaws, as well as in relative lengths of limbs and digits.

A visit to the zoo would have told anyone as much. However there was more. To ram home the point, Allan Wilson joined forces with a biochemist, Lorraine Cherry, and a herpetologist (an expert on reptiles and amphibians), Susan Case. They did their own exercise in taxonomy by comparing the difference between chimpanzees and humans over nine body shape dimensions that had served well to classify taxonomically frogs and toads—the order *Anura*—one of the most diverse groups of vertebrates in the animal kingdom. These dimensions included head length, forearm length, and eye-to-nostril distance. By these criteria, they reported, the difference in body shape between humans and chimpanzees was greater than that between the two most dissimilar frogs studied—frogs so unalike that they had been put into separate taxonomic sub-orders! We can appreciate how wide the gulf was by reminding ourselves that taxonomy groups similar species into a genus, similar genera into a family and similar families into an order. So the differences between chimps and humans were greater than two dissimilar families of *Anura*—midwife toads and tree frogs!

> Humans and chimpanzees also differ significantly in many other anatomical respects, to the extent that nearly every bone in the body of a chimpanzee is readily distinguishable in shape or size from its human counterpart. Associated with these anatomical differences there are, of course, major differences in posture, locomotion, methods of procuring food, and means of communication. Because of these major differences in anatomy and way of life, biologists place the two species not just in separate genera but in separate families. So it appears that molecular and organismal methods of evaluating the chimpanzee–human difference yield quite different conclusions.

This is how stark the disparity was. Molecular comparisons between humans and chimps compelled scientists to place them in the same genus, morphological comparison compelled them to push them so widely apart they ended up in separate sub-orders! Creationists love this paper. This is because they stop reading at this head-scratching part where two scientific methods appear irreconcilable. Science has got itself into a mess, they say, perhaps chimpanzees and humans are not related after all, perhaps a Divine solution makes more sense?

But, for those careful enough to read on, King and Wilson present a way out of the thicket. Since it appeared, they said, that molecular change had accumulated in the two lineages at approximately equal rates, despite a striking difference in rates of morphological evolution, the major adaptive shift which took place in the human lineage was probably not accompanied by accelerated protein or DNA evolution. They suggested that the differences between the two species were less to do with the classical evolutionary mechanism of mutation and selection but lay in the activity of genes—how much protein they made—and the timing of that activity. The phenomenon is called gene expression:

> According to this hypothesis, small differences in the time of activation or in the level of activity of a single gene could in principle influence considerably the systems controlling embryonic development.

Could it really be that subtle differences in the timing and amount of expression—active protein production—of a very similar, or even identical, genetic recipe could make a chimpanzee, on the one hand, and a human on the other? In a look to the future, King and Wilson concluded:

> Biologists are still a long way from understanding gene regulation in mammals, and only a few cases of regulatory mutations are now known. New techniques for detecting regulatory differences at the molecular level are required in order to test the hypothesis that organismal differences between individuals, populations, or species result mainly from regulatory

differences. When the regulation of gene expression during embryonic development is more fully understood, molecular biology will contribute more significantly to our understanding of the evolution of whole organisms. Most important for the future study of human evolution would be the demonstration of differences between apes and humans in the timing of gene expression during development, particularly during the development of adaptively crucial organ systems such as the brain.

Those new techniques have now arrived and, over the last decade, several research groups have been discovering just how right King and Wilson were.

First of all, let there be no doubt about the degree of similarity between the two genomes. I calculated the average of 33 estimates of sequence identity that had appeared in the literature to the end of 2003 and came up with a figure of 98.51% identity between chimpanzee and human, leaving 1.49% sequence difference, slightly closer than the much-bandied-about figure of 1.6%.

In April 2002, a genome research team led by Wolfgang Enard, Philipp Khaitovich, and Svante Paabo, from the Institute for Evolutionary Anthropology in Leipzig, took the first step toward verifying King's and Wilson's prediction. They reported a comparison of variation in gene expression within several primate species, and between them. They looked at the activity of genes from three different organs of the body—blood, liver, and brain neocortex—by measuring the amount of messenger RNA (the intermediate, in transcription, between DNA and the corresponding protein) produced by each of the genes. They used an ultra-modern technique called gene chip technology to compare the amount of messenger RNA produced by each organ for each species with a reference set of 18,000 human genes.

For both blood and liver, gene expression between chimps and humans was very similar—and much different from the macaque— fully reflecting their evolutionary proximity. However, they reported a 'stark contrast' for the neocortex of the brain. Here the chimpanzee expression patterns were much more similar to the macaque than the humans, thanks to a dramatic five-and-a-half-fold acceleration in the rate of change in gene expression levels in the lineage leading to humans. In other words, gene expression in human brains had shot ahead of that in chimps.

Knowledge that changes in gene expression patterns in the human brain had been rampant was one thing—now the job was to find out what mechanisms had led to those expression differences. In 2003, Mario Caceres, then at the Salk Institute, and a team that included Dan Geschwind from UCLA and Todd Preuss from Emory University in Atlanta, produced some fascinating support for Enard's and Paabo's results a year earlier. They reckon that changes in gene expression levels are *the* single most important distinction between human and non-human primate brains. They compared the patterns of gene expression in brain, liver, and heart and identified 169 genes that showed changes in expression between human and chimps. Over 90% of the brain genes had been up-regulated in human brains—they had higher levels of gene expression—whereas in human and chimpanzee heart and liver equal numbers of genes had been up-regulated and down-regulated. Although they parted company slightly with Enard and Paabo in that they found that the total number of genes differentially expressed in liver and heart was higher than in brain, only the brain had this enormous bias toward up-regulation. The identity of these up-regulated genes gives us a clear idea of what processes, in the brain, evolution was concentrated on. Most were associated with neuronal activity, particularly the speed of transmission of nerve impulses across the synapses joining neurons together. Others were associated with energy production, for instance, the transport of a sugar, lactate, as an energy source to neurons. All this supported the idea that human brains are much more energetic, gram for gram, than the brains of other animals. There were also genes associated with protecting body cells from harmful chemicals, and genes called protein chaperones, which, as their name implies, protect proteins from all sorts of abnormal aggregation and folding that are common features of such neuro-degenerative conditions as Alzheimer's disease. A picture was emerging of a brain evolved to operate faster, consuming more energy, with inbuilt protection for a longer life-span than is normal for the rest of the primates. Bigger, faster, greedier, longer-living— that's the evolutionary story of the human brain.

Most of the studies comparing rates of gene expression are very careful to point out one very potent potential confounder of their results. Promoter or regulator genes are like the accelerator and brake on a car. They can react swiftly to a change in the environment and tone down, or speed up, the transcription of a gene under their control. So a large proportion of the gene expression differences between species, for instance, could have reflected some temporary reaction to some environmental stress—a good or bad hair day in the species concerned, rather than some long-term inbuilt modulation of that gene's activity. But herein lies the whole point about the economy of evolution. It may make much more sense for evolution to favour changes in the regulation of a gene rather than sequences within the gene itself—permanent fixes that cannot easily be undone. It is a much more sensitive and faster way to get genes to react to the environment than amino-acid coding mutations. And what better example of an organ that must adapt quickly to changes in the environment than the brain? Study after study has found that the brain is the most conservative organ in the body in terms of natural selection acting on brain-building and brain-wiring genes. This makes sense. Brains are built on the principle of bricolage—building bits onto existing structures. The ancestry of human brains goes back hundreds of millions of years. There is simply too much possibility, in the complexity of a working brain, of undoing something that is working quite well. As a general rule it makes sense to conserve genes but alter their expression—keep the accelerator pedal but 'press it to the metal'. This is precisely what a team of scientists connected with Gregory Wray's laboratory at Duke University found with a gene called *prodynorphin*. In selecting *prodynorphin*, Wray and his colleagues backed a hunch:

> We focused on the prodynorphin gene because it has been shown to play a central role in so many interesting processes in the brain. These include a person's sense of how well they feel about themselves, their memory and their perception of pain. And it's known that people who don't make enough prodynorphin are vulnerable to drug addiction, schizophrenia, bipolar disorders and a form of epilepsy. So, we reasoned that humans might uniquely need to make more of this substance, perhaps because our brains are bigger, or because they function differently.

Prodynorphin is the precursor of endorphins. These are the body's natural opiates and are still popularly thought to be responsible for the 'runner's high'—the feeling of mild exhilaration we get from strenuous exercise (though recent scientific work has implicated another chemical.) They not only regulate the perception of pain, but its anticipation, and they are also important factors in social attachment and bonding. The part of the gene that actually codes for the protein is one of the most highly conserved throughout the primates—it shows no variation at all in humans or even between humans and any of the great apes. However, their overt premise was a long-overdue and explicit test of King's and Wilson's idea—that any variation with the gene was likely to be in its regulation:

> Thirty years have passed since King and Wilson argued that human evolution owes more to changes in gene regulation than to changes in gene structure, and although their theoretical justifications remain strong, empirical study of human regulatory evolution has not kept pace.

Their research was only the second ever study to test for evolution of DNA sequences that regulate gene expression by recruiting protein molecules which either initiate or block transcription. To test this they sequenced 3,000 bases of the regulatory DNA upstream of the coding, or transcriptional, segment of the *prodynorphin* (PDYN) gene from humans and seven other species of primate. This contains a 68 base-pair 'tandem repeat polymorphism'—a section of variable DNA which can exist as one copy or several copies, aligned end-to-end. This is the key promoter sequence which ramps up the activity of the gene. In humans alone they discovered no less than five nucleotide substitutions—far more than you would expect from chance or neutral evolution in such a short sequence. Statistical genetic tests allowed them to say, with high certainty, that this was the result of strong accelerated evolution peculiar to humans. In contrast, there was evidence that in the actual *PDYN* gene downstream of the promoter sequence there had actually been negative or purifying selection operating—actively conserving the original sequence structure.

Although the 68-base-pair promoter only exists as a single copy throughout the primates, there are different alleles (versions) of the gene in human populations, such that some individuals only carry one copy of the promoter sequence, others two, three, or even four. These are called repeats. The one- and four-repeat variants were very rare in every human population they looked at, but the two- and three-repeat variants differed dramatically between populations. The three-repeat was rare in China and Papua New Guinea, at less than 10%, intermediate in Cameroon and India, at around 35%, and highest in Ethiopia and Italy at over 60%. In contrast, the two-repeat had a moderate frequency of over 30% in Ethiopia and Italy, high frequency in Cameroon and India, and was approaching fixation (presence in all individuals) in China and Papua New Guinea, with frequencies of 88% and 98% respectively.

PDYN had clearly undergone a very complex evolutionary history—and all of it specific to hominids after the split from the common ancestor. More ancient positive selection had favoured the build up of five mutations within the tiny promoter sequence, and further selection had produced a family of promoter sequences ranging from one to four repeats. More recent selection, during the spread of *Homo sapiens* around the globe, had favoured one type of repeat over another in different populations.

Since the more repeats you have means the more the gene is up-regulated, what might have been the selection pressure that led to higher endorphin production in one geographical area over another? The authors are flummoxed. Apart from differentially up-regulating the gene, the specific form of regulation dealt with here is called transcriptional inducibility—a rapid response to rapid changes in the environment. A quick dab of the accelerator. But in what precise evolutionary, perhaps social, context could increased regulatory control of endorphin, affecting perception, emotion, pain, memory, learning, and social bonding, be favoured? Wray and his colleagues are vague, and leave only one faint clue. Several research groups have noticed that changes in the expression of such chemical messengers in the nervous system seem to have accompanied the domestication of dogs from wolves, and in experiments to domesticate wild animals in real time; a profound and interesting clue to human evolution to which we will return in chapter 11. For the time being the authors leave us with a resounding positive test of a thirty-year old theory:

> In keeping with the predictions of King and Wilson, our data imply that minor changes in gene regulation played a significant role in the evolution of the traits that make us human.

By 2007, evidence had accumulated for about 300 genes whose expression had undergone regulatory changes. Most of them had pepped up the rate at which genes produce protein and most of them had been exclusive to humans and the human brain. In late 2006, Mario Caceres, Todd Preuss and colleagues reported that the genes for a family of glyco-proteins (part protein, part carbohydrate) called thrombospondins had been up-regulated in the human brain. Specifically, they noticed a sixfold increase in production of the thrombospondin THSP4, and a twofold increase in THSP2 production in the cortex when they compared humans to chimpanzees. They had been led to thrombospondin by Ben Barres, of Stanford University, who reported that neurons in the brain would not form synapses—the chemical junctions between neurons—if a special kind of cell, an astrocyte, was not present. Astrocytes make up nearly half the cells in the human brain, but up until then, nobody had figured out what their function was. They are a star-shaped type of glial cell. Glial cells surround neurons in the brain, and out-number them 10 to 1; they are the glue of the central nervous system. They support neurons, insulate them from each other, and nourish them. Some of them are also involved in nerve transmission. There is a much higher ratio of glial cells to neurons in human brains compared to chimps. Somehow, the astrocytes, which do not form synapses themselves, trigger synapse production between neurons—and they do it by secreting large amounts of thrombospondin.

Surrounding the neurons and glial cells is neuropil, the basic constituent of the grey matter of the brain, and, according to neuroscientist Josef Spacek, one of the most highly organized structures in the universe: it is a huge network of fibres and synapse junctions coming from both neurons and astrocytes. This synapse-rich neuropil contains much more thrombospondin in human forebrains than it does in chimps.

About 80% of the neurons in cortex are pyramidal neurons—they are the main output units of the cortex. They have one dendrite at their apex and a number of dendrites fanning out from their base. They are bound together in large vertical arrays called mini-columns—in humans these columns can contain over 100 neurons. Katerina Semendeferi, at the University of California, San Diego, has found that these mini-columns are thicker in parts of the human brain than their counterparts in the great apes, and have larger

neuropil space around their circumference to allow for richer synaptic connections. One of the brain areas where these mini-columns
exhibit large neuropil is the most anterior part of the frontal cortex, known to play a major role in memory retrieval and executive
function—very high-order brain processes that allow us humans
to be socially competent and act appropriately in social situations,
and plan our lives in prudent and adaptive ways. This is also where
Caceres and Preuss found particularly high levels of expression of
thrombospondin. The human brain is extremely plastic and can
change the patterns and richness of synapse connections between
neurons as we learn from novel experiences or should our brain
get accidentally damaged. Increased thrombospondin levels could
enhance these plastic changes in the adult human cortex by making
our cortical networks denser, more complex, and capable of running
faster.

There is, however, a possible down-side. Thrombospondin 4 has
been shown to bind to the B-amyloid tangled plaques we see in
patients with Alzheimer's disease. They may cause some form of
inflammatory response here. All this suggests that evolution of
thrombospondin expression may have resulted in changes in synaptic organization and plasticity, and could have contributed to the
distinctive cognitive abilities that make us human, but at the cost of
our increased vulnerability to neuro-degenerative diseases. Live fast,
pay later.

In August 2007, Ralph Haygood and colleagues in the laboratory of Gregory Wray, the researcher who led the discovery of the evolution of the *prodynorphin* gene, began to tie the
whole regulation-evolution scene together. Haygood had surveyed
the promoter regions of 6,280 genes that occur in humans,
chimps, and macaques, and found that the promoters of 575 of
those genes had evolved in humans in ways not found in the
other two primates. Many of these genes were involved in the
brain—specifically in the development of neurons and carbohydrate

metabolism. Their aim was to discover just how strong that selection had been.

To do it, they compared rates of evolution between the promoter regions for all these genes and nearby intronic sequences. An intron is a length of DNA sequence close to the exon, or coding part of a gene, which does not actually code for anything itself—it is rather rudely called junk DNA. When the DNA is transcribed into messenger RNA the introns are spliced out before the rest of the molecule goes on to act as the template for a protein. Because they are rarely functional, introns tend to accumulate more mutations than the coding part of the gene—where mutation is less tolerated because there is far more chance that it will be deleterious than adaptive. The mutations that occur in introns stand less chance of being weeded out—they just pile up over time. So, if the promoter region out-performed the intron, in terms of its accumulation of mutations, it really must have been the target of very strong natural selection.

Out of the 575 sites that showed evidence for positive selection there were 250 particularly strong candidates. Since the 6,000-plus genes they had sampled constitutes approximately one-third of all human genes, they extrapolated to suggest that the promoter regions of at least 750 genes have experienced positive selection. This shows just how astonishingly prevalent evolution of gene regulation has been. What were all the genes associated with these positively selected promoter regions doing? By consulting a number of databases for their 100 highest-flying genes they found that they were overwhelmingly involved in neural development and function, belonging to categories like neurogenesis, synaptic transmission, and axon guidance (first design your network, then join it up, then run signals along it).

Chimpanzees have evolved their own style of gene regulation. Very few of the many genes involved in neural system development or the metabolism of carbohydrates show both increase in expression and positive selection in both chimps and humans. Ralph Haygood reckons that this reflects the fact that the diet of early hominids changed from the fruit-rich diet typical of chimpanzees to a much more carbohydrate-rich diet of roots and tubers. Efficient breakdown and metabolism of carbohydrates, caused by rapid evolution of key metabolism genes, would have provided the vast amounts of energy needed for brain expansion and to fuel a bigger brain.

A pattern is emerging, Haygood suggests. You might expect—given that some of the main ways we differ from chimps are our cognition, behaviour, and nutrition—that many genes involved in these areas would have undergone strong selection. But this is not so. Where numbers are reported, they tend to be more involved with things like immunity, sperm production, and smell. This observation is what led Haygood to look in the regulatory sequences of genes for the hand of evolution, rather than the genes themselves. His research is supported by a number of other recent studies which all conclude that the rate of evolution of genes in the brain of humans is, if anything, lower than that in chimps. Haygood reckons that the increasing size and complexity of the human brain constrains evolution of genes because the bigger and more complicated the works, the more damage the proverbial spanner (mutation) is likely to do when thrown into them. The answer—particularly in humans—to the dual problems of rapid response and rapidly increasing complexity has been in the evolution of powerful networks of gene *regulation*. Raising the levels of gene expression in humans amplifies the very small differences in gene coding sequence between humans and chimps, and, because many of the affected genes are transcription factors, which, in turn, control whole networks of genes, those differences can be amplified still further and made to respond sensitively to rapid environmental change. Many of these regulatory networks are housed, uniquely, in the human brain. It has taken some 35 years, but King and Wilson appear to be fully supported.

Allan Wilson died, prematurely, at the age of 55. But his scientific partner all those years ago, Mary-Claire King, now a senior professor at Washington University, is quoted giving Haygood, and everything his research represents, a vigorous nod of approval:

What a nice paper. Allan must be smiling from heaven!

CHAPTER 5

Less is More

Mary-Claire King's and Allan Wilson's prescient suggestion was that changes in the activity levels of genes—changes in gene expression—are more important than single point mutations in the DNA code that change the amino-acids in proteins when it comes to explaining the many differences between chimpanzees and humans. But is it enough? In among all the excitement over the recent research that provides resounding support for King's and Wilson's seminal paper, there was one note of caution. It came from Ajit Varki, professor of glycobiology at the University of California, San Diego. Gene expression differences between humans and chimps were far from the whole story, he argued: the genome is a much more weird and wonderful place, abounding with a whole menagerie of structural differences—changes to the very architecture of the genomes—which are also a potent force for evolution.

Varki knows what he is talking about. For years, he and his research group have been working on glycans—long-chain sugars which attach to protein and fat molecules and coat the outer surface of all cell types in the body, where they are involved in interactions or signalling between cells. Like bouncers at the door of a night club, they decide what chemical messengers are allowed to go in and out. These complex assemblies of molecules on the cell surface can be abused by pathogens, like viruses, which have evolved the ability to attach to the glycans as a first step to invading cells, infecting them, and causing disease. Classic examples include the invasion of the respiratory system by the influenza virus and the invasion of the blood by the malaria parasite, *Plasmodium*. Because any microscopic pathogen has a much smaller generation time than a human (or a chimp)—seconds or minutes compared to decades—and often a much higher mutation rate, it can always out-run its host in the constant evolutionary arms race that characterizes the so-called host–pathogen relationship. The glycans, at the front line of immunological defence, are under enormous evolutionary pressure to change and evolve in a constant attempt to stop rapidly evolving pathogens sticking to them. Despite this furious game of evolutionary roulette with pathogens, the array of glycans on the cell surface is very specific to the type of cell and glycan arrays are also highly species-specific.

The way that these glycans are produced is under close genetic control, and occasional mutations in these glycosylation genes can even completely omit a particular glycan from the assembly. Varki's study of evolution in this glycan family, and of other cell-surface molecules that form complexes with the glycans, has produced a fascinating story of chimp–human differences in which a key mutation some 2–3 million years ago has changed the way our immune systems work, compared to chimps, and may also have affected the evolution of the human brain. This is not a story of change in gene expression, but the knocking out of a gene altogether by a mutation which turned it into a pseudogene—a functionless relic. It turns out that this is one of over 200 examples of 'loss of function' mutations which have resulted in humans losing some gene function compared to chimpanzees, rather than gaining. Varki, and his friend and colleague from the University of Washington, Maynard Olsen, call this the 'less is more' hypothesis. It is a highly intriguing idea because, to the extent to which gene loss proves significant, in the greater picture of things, the 'less is more' hypothesis suggests we are actually degenerate apes rather than super-apes.

These loss-of-function mutations come about when crucial parts of the coding sequence of a gene are deleted, rendering the gene useless. In the vast majority of cases the deletion of important parts of the DNA recipe for a protein proves disastrous: the mutation is highly deleterious and the individual harbouring it swiftly perishes. However, very occasionally, the creation of a pseudogene can buck this trend and provide an opportunity for evolution in the genome. Varki had stumbled over just such an example.

Varki's research, first reported in the late nineties, concerns a specific group of sugars called sialic acids, which are found in conjunction with the range of cell-surface glycans. One of the commonest sialic acids is N-glycolylneuraminic acid (Neu5Gc for short). It is formed from its precursor, the sialic acid N-acetylneuraminic acid (hereafter Neu5Ac), by the addition of OH (hydroxyl) groups in a reaction catalysed by an enzyme called a hydroxylase. In humans this enzyme has lost all its activity with the result that we only make the tiniest amounts of Neu5Gc—and that only in embryonic and cancerous tissues. In all other tissues in our body Neu5Gc has vanished. Instead, our cell surfaces are coated with the precursor molecule, Neu5Ac. In chimpanzees, and the other great apes, the final conversion to Neu5Gc is unaffected. They have high levels of Neu5Gc in all tissues, except, interestingly, in their brains, where the concentration of Neu5Gc is much lower. However, by failing to make any Neu5Gc whatsoever, humans have no Neu5Gc in their brains at all. Varki has a feeling this difference between low levels and complete absence of Neu5Gc in the brain might prove crucial for human brain evolution, but, to date, he has not been able to come up with a plausible reason.

The inactivation of this hydroxylase is caused by a mutation of the *CMAH* gene which codes for it. A large 92-base-pair segment in the coding region of this gene has simply been deleted. This is what is known as a frame-shift mutation in the genetic code which, in effect, scrambles the message at this point—the enzyme cannot be formed. When did the mutation occur? Together with Svante Paabo, of the Institute of Evolutionary Anthropology in Leipzig, Varki investigated the long bones of some Neanderthal specimens. Although much genetic material has been lost over 30,000 years of fossilization, making DNA recovery a laborious and highly specialized business, the sialic acids survived in sufficient quantities to allow them to discover that we and Neanderthals must have shared the same mutation. Neanderthals did not express Neu5Gc in their bone marrow. Further genomic calculations put the mutation at approximately between 2 and 3 million years ago, well before *Homo sapiens* had evolved. This is a drastic mutation at an archaic time in hominin evolution, possibly around the time of the emergence, in Africa, of *Homo erectus*. Since this was after the evolution of bipedalism, but before the major period of brain expansion, it could be that the mutation produced a selective advantage that was later employed in brain evolution, or it may have nothing, per se, to do with brain evolution at all. In which case, what might be the alternative?

One of the main immunological differences between us and chimps is that chimps do not appear to be susceptible to the common malaria parasite, *Plasmodium falciparum*, which kills over 2 million humans every year and affects 500 million. More than 80% of those deaths occur in Africa, south of the Sahara. Varki's group looked at the relationships between *Plasmodium falciparum* and the sialic acids which coat red blood cells—the target for the parasite. *P. falciparum* showed no affinity at all for Neu5Gc, the sialic acid common in chimps: it cannot therefore use this sialic acid as a Trojan horse to get into red blood cells—hence the chimps' immunity. However, chimpanzees do suffer infection from a closely related species of Plasmodium, *P. reichenowi*, which strongly favours Neu5Gc. On the other hand, *P. reichenowi* has no affinity for Neu5Ac and consequently, today, does not infect humans. The researchers suggest that the original loss of Neu5Gc in hominins, about 2 million-plus years ago, was a drastic evolutionary measure to defend against malaria, which, in those days, was caused by *P. reichenowi* and infected hominins and apes alike. It is still a mystery why apes did not similarly lose Neu5Gc. Varki thinks that the *CMAH* deletion was a chance event in early hominids that simply did not occur in apes, or their ancestors. There is very little information on *P. reichenowi* virulence in chimpanzees today and it is possible that they have always somewhat tolerated *P. reichenowi*, such that there was less selection pressure for a resistance mechanism to evolve. Eventually, Varki theorizes, *Plasmodium* fought back against hominins via the evolution of *P. falciparum* from a strain of *P. reichenowi*, such that malaria continues to torment us to this day. Research is currently going on to test this hypothesis.

This drastic replacement of sialic acids at the cell surface, possibly driven by the need to survive parasitic attack, has resulted in some other very profound knock-on effects in the human immune system, further differentiating us from chimps and other primates. In fact, Varki's research seriously challenges the rationale that led to chimpanzees being used as animal models for human disease because of their genetic proximity to us. In fact, says Varki, many human diseases are different, either in incidence or severity, because of these evolved differences in the genes that underpin our immune systems. These further effects involve molecules called Siglecs (short for sialic acid recognizing lectins) which are part of a huge family of antibodies called immunoglobulins. Siglecs straddle the cell membrane, with their tails jutting into the cytoplasm on the inside, and their heads binding to sialic acids on the outside. They thus form part of the rich and complex Siglec–sialic acid–glycan system of the cell surface which recognizes 'self' from 'non-self' and organizes immune responses accordingly. Differences between humans and chimps with respect to these Siglecs may help to explain why we humans suffer from AIDS and chronic hepatitis whereas chimps don't.

Most of the Siglecs belong to a single group called CD33. They act to dampen down the immune response to antigens, such as the foreign proteins coating the surface of invading bacteria and viruses, thus preventing an overreaction of the immune system which, in some cases, can be more harmful than the infection itself. Most of our family of immune cells, like granulocytes and macrophages, have arrays of Siglecs on their surface. However, one striking exception are human T-cells, which, unlike chimps and the other great apes, have no Siglecs on their surface whatsoever. Our aberrant human T-cells are just one of a string of rapidly evolving changes to the human Siglec line-up that followed from that CMAH gene deletion 3 million years ago which left us without Neu5Gc.

The lack of regulatory Siglecs on human T-cells makes them more reactive, compared to those of the great apes. Varki considers that this may explain the frequency and severity of T-cell mediated disease in humans. In particular, chimpanzee CD4+ T-cells (so-called helper T-cells) are rich in Siglec 5, which is absent in humans. These cells are implicated in human diseases including AIDS, chronic active hepatitis, inflammatory bowel disease, rheumatoid arthritis, type 1 diabetes, multiple sclerosis, and psoriasis. There are also differences in Siglec expression on B-cell lymphocytes which may leave humans open to lupus erythematosis—which has not been reported in chimps. All these are examples of auto-immune disease where the human body attacks itself.

Don't tell Navneet Modi, David Oakley, Nino Abdelhady, and Ryan Wilson that primate immune systems are so similar to our own that drugs previously tested on monkeys are safe for human volunteers. On 12 March 2006, they turned up at a special unit in Northwick Park hospital in north London to take part in the trial of a new monoclonal antibody—TGN1412—designed to fight some cancers, and auto-immune diseases like rheumatoid arthritis. Monoclonal antibodies are antibodies derived in large quantities from a single cell line for use against a specific foreign protein, or antigen. TGN1412 is a powerful immune-system stimulant, designed to override the way the body normally regulates its immune response, by attaching to T-cell receptors and interfering with their regulatory function, thereby turbo-charging T-cell immune response—taking the brakes off. The dose these human guinea pigs were given was diluted 500 times from the dose that had previously been administered to monkeys with no ill-effects. Ninety minutes after having received the injections they began suffering from violent headaches, nausea, diarrhoea, and massive swellings of the head and neck. Their eyes turned yellow. They were described by fraught wives and girlfriends as looking 'like the elephant man', and one of them described himself as looking like 'toxic waste'. Several months afterwards, Simon Hattenstone, for the Guardian newspaper, interviewed them. David Oakley:

> 'They had to pull the curtain around Nav. He started throwing up first. I was one of the last, but when I threw up, I threw up big. I probably brought up over a litre of bile. I didn't know you had that much bile in your stomach. A whole container of green bile—I thought, "Whoa! What the heck is that?" The nurse looked at me and went, "Bile." Even she was shocked by how much I'd brought up.'

> Katrina, Oakley's partner, was told that Oakley had suffered a reaction, but nothing more. 'She got to the hospital at 4 a.m. It was gloomy and dark, and she was wandering around, hysterical,' Oakley says. 'Finally, a doctor found her and pointed her to intensive care. One of the doctors explained that we were on machines and we had tubes going in and were really swollen. They prepared her for the worst because we weren't a pretty sight. They said they were doing their best, but they didn't know what the chances of surviving were.' Despite the warning, Katrina was shocked by his appearance. 'She said my head looked like a ball and my eyes were like slits. My stomach was completely swollen. She thought I had my hands over my stomach till she saw them over the side. I was shivering, my

whole body vibrating. I had a temperature of 41/42 degrees—
43/44, you're dead.'

These men had experienced violent immune overreactions. Their
bodies began producing cytokines, immune system signalling molecules which mobilize T-cells and macrophages and direct them to the
site of infection. But in this case, positive feedback ran amok. More
and more cytokines were released—the so-called cytokine storm—
dispatching waves of immune-system cells to all the major organs of
the body, where they turned on the trial members, causing damage to
hearts, lungs, livers, and kidneys—massive multi-organ failure. The
six volunteers were eventually rushed to intensive care and stabilized
with steroids. David Oakley has now contracted a vicious form of
lymphoma, while Ryan Wilson had to have several fingers and all his
toes amputated—they had turned black with necrosis. Most of them
will never fully recover and have been left susceptible to disease for
the rest of their lives.

In the wake of the incident, the Medicines and Healthcare Products
Regulatory Agency (the MHRA) found that 'an unexpected biological
effect' was the most likely cause of the incident, and the Duff Report
that followed, commissioned by the then Health Secretary, Patricia Hewitt, suggested caution when testing novel immune-system
drugs—especially antibodies. Ajit Varki thinks he knows exactly
what that biological effect was. The lack of CD33 Siglec expression in
humans may have directly contributed to helper T-cell hyperactivity,
releasing the cytokine storm that nearly killed them. Varki asked
TeGenero, the company that had developed TGN1412, if they would
donate a sample of the antibody so that he could test it against chimpanzee blood. The company declined and so he was unable to test his
theory that because human T-cells are already primed to overreact—
perhaps stripped of their Siglec brakes because of a massive selection
pressure to defeat malaria millions of years ago—the drug company
researchers' further stimulation of the human immune system was

akin to slapping a thoroughbred stallion firmly on the rump, causing it to bolt.

Today, medical researchers are frantically trying to develop a vaccine against malaria, which has proved to be a truculent and obdurate foe. The American malaria vaccine pioneer, Dr Rip Ballou, experienced the cytokine storm at first hand a few years ago, having volunteered to be injected with a trial vaccine after deliberately exposing his forearm to malaria-infected mosquitoes. The vaccine was a dud but his immune system wasn't. His garden began to swim before his eyes during a burger lunch with friends and later he collapsed in the middle of giving a scientific talk in San Diego. Millions of people in the tropics depend on this human cytokine reaction for some degree of protection against the malaria parasite—it is called natural immunity and it is what gives you the feverish temperatures and cold sweats which are the common symptoms of infection. The problem with the cytokine reaction is that it is so unpleasant and does so much collateral damage to your own body that tiny babies and young children often fail to survive a malaria infection and the mobilization of the body's defences to fight it. If you can get past childhood, natural immunity gradually builds up and lessens the severity of subsequent infections. Natural immunity will cure you, after a fashion, if it doesn't kill you first. Hair-trigger T-cells were evolution's quick and dirty fix for violent malarial insults millions of years back among our hominin ancestors.

In their 12 August 2006 issue, the *Toronto Globe and Mail* printed the story of a local man, Ron Rosenes, who had lived with HIV for 25 years:

> He has always suspected something more than good fortune shielded him from the worst of the disease. Mr. Rosenes feels he contracted the AIDS virus before anyone knew it existed and 'safe sex' became the mantra of a generation. 'I was sexually very active in the late seventies,' he said. He has lost several friends and a close cousin to AIDS. In 1990, he watched his

partner of 15 years die in the home they shared. But even as
he lost weight, left his job and made do with the 'sub-optimal
therapies' of the early nineties, Mr. Rosenes held on.

He never suffered the opportunistic infections others
endured. He never picked up anything life threatening.
Although his viral load soared and certain immune cells plum-
meted, blood test after blood test showed a healthy presence of
other immune cells.

'It became more obvious to me, especially after [my partner]
died, that I was being spared.'

The 59-year-old Toronto man may well have his family history to thank. Mr. Rosenes carries a gift in his genes—a mutation that confers a natural resistance to HIV. In all likelihood, it
was passed down to him from his ancestors in Europe, where
the protective trait is most prevalent.

Mr. Rosenes' 'gift' was one copy of a mutation of a gene called CCR5.
Whereas the story of the sialic acids and the Siglecs is a tale of relatively ancient human-specific evolution, the case of the CCR5 gene
is modern. The gene codes for a receptor molecule which binds to
macrophages and helper T-cells and activates and mobilizes these
white blood cells. A few years ago it was found that HIV has a way
to bind to this receptor, using it as a Trojan horse to gain entry into
the cell. The mutation in CCR5 led to loss of function of the gene,
vastly reducing the amounts of receptor molecule on the cell surface
and therefore depriving HIV of its foothold. The macrophages were
much less sensitive to HIV and much less of it got into the cell. The
genetic cause is a classic 'less is more' deletion of 32 base-pairs of
the coding portion of the gene—rendering it useless. The mutated
allele is therefore called CCR5 del 32. It is principally found in Europe
and Western Asia, where approximately 10% of all people have one
copy of it and 1% have two. It is less common in northern Africa and
occurs nowhere else in the world. Those small numbers of individuals who have two copies of del 32 are completely immune to HIV
and even those who have one copy have much lower viral loads in

their bodies and slow progression to AIDS by several years. There is a very interesting controversy as to when this mutation arose, whether or not it was positively selected, and what that selection pressure might have been. The controversy goes a long way to explaining just how difficult it is to reconstruct human evolutionary history even when you have at your disposal various powerful techniques to model human genetic variation among populations today. In the case of *CCR5 del* 32, obviously the original context could not have been protection against HIV—the time elapsed since the first cases of HIV-AIDS is simply not enough. What infection could have provided the context, why and when?

John Novembre, now at the University of Chicago, Alison Galvani, and Montgomery Slatkin have looked at the geographical distribution of the *del* 32 mutation, the evidence for the selection pressures that might underpin that distribution, and have estimated the date of the mutation by a variety of techniques to be between 700 years and 3,500 years ago.

CCR5 del 32 is not spread evenly throughout European populations. It is at highest frequency, 16%, in northern Europe and declines to 6% frequency in Italy and 4% in Greece. The largest area of high frequency takes in a broad swathe of northern Europe, including Sweden, Finland, Belarus, Estonia, and Lithuania, with additional peaks on the northern coast of France, parts of Russia, and Iceland. This has led to one suggestion that it arose in Scandinavia before the Vikings populated Iceland 1,000–1,200 years ago and that gangs of marauding Vikings then subsequently spread it further around Europe, west to Iceland and east to Russia.

A number of researchers have tried to put molecular genetics, microbiology, and history together into a pleasing story. All have jumped onto the '700 years old' date and this has quickly led them to the waves of Black Death pandemics that swept Europe during the Middle Ages, beginning around 1348. The leading theory for many years has been that the Black Death was bubonic plague caused by the bacterium *Yersinia pestis*, carried by rats. The pandemic that started in 1348 is thought to have killed 30% of Europeans and the Great Plague of 1665 killed 15 to 20%. However, one American research group compared control laboratory mice with mice that had the *del 32* gene 'knocked in' to them. They failed to find any difference in infection when both were injected with *Yersinia pestis*, suggesting either that *del 32* did not arise in response to *Yersinia pestis* pandemics or that *Yersinia* was not, after all, the infective agent in Black Death. Research by Alan Cooper, then of the Ancient Biomolecules Centre at Oxford University, suggests the latter because he failed to find any signs of *Yersinia pestis* infection in skeletons dug up from plague pits in East Smithfield in central London. The most impressive put-down of the *Yersinia* theory comes from Sam Cohn and Lawrence Weaver, from Glasgow University, who point out that bubonic plague—driven by flea-infested rats—is a disease much more common in the tropics, sub-tropics, and far East (where the *del 32* allele is virtually absent) than it is in Europe, particularly northern Europe, where the frequency of the *del 32* allele is highest. Further, no bubonic plague for which they can find records ever managed to kill more than about 3% of the population. *Yersinia pestis* is a disease involving fleas, a bacterium, and rats, they say, in which humans only occasionally get caught in the cross-fire.

Professor Christopher Duncan and Dr Susan Scott, from the University of Liverpool's School of Biological Sciences, were equally unimpressed with the *Yersinia* theory. They looked for a disease that could spread like wildfire and had close to 100% mortality. They also noted in accounts of the plague that visited Athens in 430 BC, and France in the fourteenth century, that there appeared to be two sets of symptoms. Only those who died after about five days had the pustules and buboes associated with the Black Death; victims who succumbed sooner had massive bleeding from the nose and in vomit, which suggested some haemorrhagic fever similar to viral diseases like Ebola and Lassa fever. However, Cohn and Weaver find fault with the Liverpool analysis also. They take umbrage at Duncan and Scott's assertion that the plagues were confined to Europe, stating firmly that the Black Death arose outside Europe, in Egypt, India, China, and central Asia. North Africa and Asia Minor were devastated far more comprehensively by Black Death than the whole of Europe—which is the almost exclusive preserve today of the *del* 32 mutation.

Duncan and Scott had used computer models to demonstrate how haemorrhagic fever could have provided the selection pressure that forced up the frequency of the *del* 32 mutation from one in 20,000 at the time of the Black Death to values today of one in ten. Small and large epidemics over centuries, they have argued, have nursed the *del* 32 mutation to the very high frequencies we now see. But, for Cohn and Weaver, Duncan and Scott appear not to have read their medical history properly.

Neither, according to Cohn and Weaver, did Galvani and Novembre, who in 2005 reported that their data favoured smallpox as the culprit. Because smallpox epidemics occurred frequently, they argued, young children were the only immunologically naive individuals in a population, the only age-band having no immunity, making smallpox a childhood disease. Most Europeans, they say, were infected by smallpox before the age of 10 and the fatality rate was about 30%. So, not only did smallpox kill more individuals, it disproportionately hit the young and therefore killed off more reproductive potential. This would indeed have generated sufficient selection pressure. It also fits the molecular evidence better since both HIV and smallpox are viruses known to interact with the CCR5 receptor, whereas plague is caused by a bacterium and most evidence suggests it does not recognize CCR5. Galvani and Slatkin mathematically modelled the strength of selection on *del* 32 and calculated that it was consistent with the selection that would have been produced by smallpox. They claim few alleles have ever been shown to have undergone a stronger selection pressure than that.

The case is well argued, but, like so much evolutionary genomics, it is based on theoretical genetics modelling, not the 'real world'. It is a mathematical simulation, if you like, of past events. Cohn and Weaver also find fault with Galvani's scholarship. Smallpox originated outside Europe, in Somalia and India, and there is no evidence, they say, that Europe suffered disproportionately to the rest of the world. Indeed, when Europe was reeling from Black Death/smallpox in the sixteenth century, the New World (lamentably free of the *del* 32 mutation today), suffered far more.

The precise pattern of spread of resistance is still a matter of debate. It may not have required Viking sea-power, it could have simply radiated throughout Europe from a base in Scandinavia, or it could have originated in central Europe and increased its frequency in a northerly direction, with declining frequency to the south. If smallpox was the real culprit, and since smallpox comes originally from cattle, there might be something in the pattern of livestock rearing and husbandry in northern, as opposed to southern, Europe that is important. It is possible that this fresh viral insult to humans might have come about because humans lived cheek by jowl with their livestock in early agrarian communities, allowing viruses to jump between the species. This is called zoonosis. But a zoonosis when?

A consortium of scientists from Boston, including Pardis Sabeti, Nick Patterson, David Reich, David Altschuler, and Eric Lander, have challenged Galvani's research. From their calculations, they conclude that there is no evidence that *del* 32 had undergone any positive selection! They used roughly the same argument they employed in their critique of Bruce Lahn's claims for recent evolution and selection of *ASPM*, which we discussed in chapter 3, and as such, their criticism must be weighed carefully—Lahn has been able to muster a typically robust counter-attack! Indeed, although the data for *del* 32 appear not to stand out when compared to data for a number of other alleles, this cannot prove that *del* 32 was not selected for—it simply cannot support the idea that it was. Sabeti and her colleagues are careful to acknowledge this when she says: 'Given the biology of the gene, it is certainly possible that it has been subject to some selection despite the lack of clear evidence.'

Sabeti also questions Galvani's estimated dates for when the *del* 32 mutation arose. Her group estimate it at more than 5,000 years ago, not 700. Their argument is strongly supported by the recent discovery of the *del* 32 mutation in DNA extracted from Bronze Age human burial sites from central Germany that are 3,000 years old. Whether or not the Black Death was involved in further selection for *del* 32, it cannot have been the event that began the process of evolution on that gene. So we are left arguing as to whether or not *del* 32 underwent selection, and, if so, what the infectious agent was, and when that selection occurred. Cohn and Weaver have questioned everyone else's microbiological candidate but have no suggestions of their own. Perhaps early animal herders, some 5,000 years ago, first contracted smallpox from domesticated cattle and the *CCR5 del* 32 mutation resulted. It may have received bursts of selection pressure ever since from viruses that interact with the CCR5 receptor, culminating in the protection it currently offers HIV-infected people who carry it.

In support of this, Stan Cohn and Lawrence Weaver do have one very interesting, but totally unexplored, suggestion. If you take the oldest estimate (Sabeti's) for the age of the *del* 32 mutation, which is 5,000 years ago, it coincides with another recent human-specific mutation, for lactose tolerance. There is overwhelming and undisputed evidence for strong positive selection for this in those human groups that herded cattle and ate dairy products. That mutation also has a European frequency gradient similar to that posited for *del* 32— increasing in frequency toward northern Europe. Stan Cohn suggests that *del* 32 ought to be looked for in a much more systematic and fine-grained way among those groups within large geographical areas of the world which are also lactose-tolerant and have a history of domesticating cattle. Presence or absence in Africa or Asia is far too broad-brush an approach. His feeling is that the solution to the mystery of the 'less is more' evolution of *del 32* will be found somewhere in the recent evolutionary history of *Homo sapiens*, after the emergence of the earliest agrarian communities of the Holocene period, 10,000 years ago. I think he is absolutely right, although for the moment, the precise provenance of Ron Rosenes' AIDS-resistant 'gift'—the *CCR5 del 32* mutation—remains up in the air.

One of the most audacious and speculative examples of 'less is more' concerns mutation of a gene, *MYH16*, which was discovered during research into muscular diseases by a team led by Hansell Stedman of the University of Pennsylvania School of Medicine. Stedman's group found a frame-shift nonsense mutation of this gene— a deletion or drop-out of a bit of genetic code that renders the gene useless—rather like the *del 32* mutation we have just discussed. *MYH16* codes for one of the protein chains that make up muscle fibres. You can imagine these protein molecules as coiled springs which provide the contractile force for the muscle. Loss of *MYH16* function has been shown to drastically affect both muscle fibre size and the overall size of certain chewing muscles which, in most primates, are large and powerful and attached to bony ridges of the skull. The team suggest that the mutation first arose approximately 2.5 million years ago and that a reduction in the size and power of these huge jaw muscles made it possible for the cranium to expand, rather like a barrage balloon lifting off once the guy ropes have been slashed. It was immediately dubbed the 'room for thought' mutation because Stedman was quick to notice the potential evolutionary implications of his medical work and pointed out that the date for the mutation corresponds rather neatly with the emergence of more gracile—more delicately built—hominids with bigger brains.

There is back-up evidence from plastic surgery research that the genetic manipulation of muscle size does have significant knock-on effects on the anatomy of bony attachment sites. Experiments have been done with primates to perfect plastic surgery on humans.

These have involved resection (partial removal) or transposition of masticatory (chewing) muscles and have led to changes in the shape of the cranium. This leads Stedman's team to reason that a sudden mutation radically reducing the contractile force generated by these muscles would have had a number of knock-on effects on cranium shape. In particular, they say, the reduction in size and force of the masticatory muscles would not require the large bony sagittal crest that runs along the top centre of the skull from back to front in most mammals and reptiles, and is seen in non-human primates as well as in Paranthropus and Australopithecines. The great fan of the temporalis muscles—the strong jaw muscles—is anchored to this sagittal crest. The zygomatic arch, which runs across the upper jaw, would also be reduced and there would be a reduction in stress across suture lines of the bony plates of the skull and the developing dura mater—the tough lining of the brain that runs directly under the skull.

To establish the age of the mutation they compared MYH16 in human, chimp, orang-utan, macaque, and dog. For most of evolutionary time synonymous mutations greatly outweighed non-synonymous mutations, suggesting that little or nothing of evolutionary significance was happening. Only in the lineage leading directly to Homo sapiens did a flip occur and the number of non-synonymous mutations increase. They assumed this rate of mutations had held constant since the chimp–human split from the common ancestor and this allowed them to estimate an age of 2.4 million years for the deletion.

Reactions to the Stedman paper, which appeared in Nature in 2005, were mixed, to say the least. Some anthropologists were intrigued but many were hostile. The major criticism was that a sudden drastic mutation to reduce chewing-muscle size, unaccompanied by a host of buttressing evolutionary changes to tooth size, jaw size, and increased brain size, would have been catastrophic, not adaptive. As it is vanishingly unlikely that such a comprehensive suite of evolutionary changes could all be made at the same time, the slack-jawed owners of the MYH16 mutation would have been at a lethal disadvantage and would have perished. And it is absolutely true that Stedman's research therefore throws up far more questions than it answers. However, at the time of going to press his work still stands—an intriguing oddity with question marks all over it—and, to date, Stedman has not taken the question of hominin skull evolution any further.

Xiaoxia Wang, Wendy Grus, and Jianzhi Zhang have reported on a systematic survey of chimp and human pseudogenes which looked for those that had been adaptive rather than deleterious. They identified 887 pseudogenes that had been formed by 'less is more' deletion of genetic sequence within existing genes, of which 80 were present in chimps as functional genes, many of which were involved either in the immune system or the sense of smell. They needed a 'proof of the pudding' case history to show the principle that creation of pseudogenes in humans could be adaptive and had been positively selected for and they focused on CASPASE12, which is involved in sepsis and the immune response to bacterial toxins. In humans, the gene had been made non-functional by one single base-pair substitution of the genetic code—enough to cut short the resulting protein chain and sabotage its function.

They tested for signals of positive selection going back 200,000 years. The non-functional, or T-allele, is fixed at 100% (everyone has two copies of it) in all world populations except sub-Saharan Africa and has a frequency of 89% in people of African descent. It results in decreased mortality from, and incidence of, severe sepsis. They calculated that somewhere between 51,000 and 55,000 years would have been necessary for the pseudogene to have reached these frequency levels. This fits with the main out-of-Africa migrations of Homo sapiens, which are thought to have occurred between 40,000 and 60,000 years ago. CASPASE12 appears to dampen down the immune response in those primates that have a working version of it. It may be that, as we have seen with our T-cells, we humans lost the function of *CASPASE12* in order to turbo-charge our immune systems against a new range of unpleasant and harmful toxins that might have cropped up in our diet and immediate environment as waves of human migration thrust us into new areas of the globe, and against which we needed to respond with more urgency.

It is clear that the strongest support for the 'less is more' hypothesis comes from the evolution of the highly aberrant human immune system. This is the signature of evolution in a hurry, erecting hasty, even reckless defences against a host of Pleistocene infections that could have otherwise put an end to the story of man almost before its opening chapters. Our human immune systems are substantially different to those of the other primates. In immunological terms we are freaks of Nature and our older generation today is paying the price for Nature's quick and dirty solutions via a range of auto-immune diseases and diseases of old age.

CHAPTER 6

More Is Better

So far we have discovered that the simplistic idea that point mutations which cause substitutions of DNA bases—the classical 'mutation followed by selection' mechanism of evolution—is an inadequate explanation of human evolution and human–chimp differences. Already we have added gene expression and the loss of function of genes as two important further mechanisms of evolutionary change. There is more to come!

In the last chapter we considered pseudogenes that have been created through deletion of DNA sequence. But there is a second way of making pseudogenes and that is to 'retrocopy' them back into the genome from their messenger RNAs. Rather than take a gene out of the genome by rendering it useless, leading to 'less is more', such 'retroposition' will make two copies of that gene—doubling its presence in the genome. Until recently, this was thought to be irrelevant because the retrocopies have left behind their regulatory sequences. When DNA is transcribed into messenger RNA as a prelude to making protein, as a rule only the exons or open reading frames of the gene—the part that actually carries the information to make the protein—get transcribed. The introns, which constitute the regulatory elements, are spliced off. So, received wisdom goes, if a retrogene is formed by the reverse transcription of only the coding portion of the gene it will lack vital regulatory elements and be as useless as a car without ignition, accelerator, and brakes. It will be a piece of scrap.

Fabien Burki and Henrik Kaessmann, from the University of Lausanne, however, have provided us with a wonderful example of how a gene can copy itself and re-insert that copy somewhere else in the genome *and* manage to avoid the fate of being stillborn. Instead it has evolved to make a huge adaptive contribution to the way the brain works. Glutamate is one of the principal chemicals in the brain and nervous system that transmits nerve impulses. It allows a succession of neurons to fire so that an impulse is carried rapidly across a neuronal network. An enzyme, glutamate dehydrogenase, is vital for a chemical step that recycles glutamate molecules so that they constantly become re-available to transmit subsequent nerve impulses. The enzyme loads the gun, if you like, that fires the nerve impulses. Glutamate dehydrogenase (GDH) exists in two almost identical forms, each coded for by closely related genes, *GLUD1* and *GLUD2*. *GLUD1* sits on chromosome 10 while its sister gene sits on the X chromosome. It is *GLUD2* that is the retrogene. It was formed from *GLUD1* when the messenger RNA of the gene was reverse-transcribed back into DNA after the intron of the gene had been spliced off. It should therefore be a non-functional relic, but, by chance, it has popped up on the X chromosome next door to some DNA sequence that it has been able to co-opt as a regulatory sequence. Burki and Kaessmann estimate the age of *GLUD2* to be about 23 million years—after the split of the great apes from the Old World monkeys, but before the major subsequent splits of the hominoid lineage, and certainly before the split of humans and chimps from the common ancestor. So, *GLUD2* is a gene common to humans, chimps, gorillas, and orang-utans only.

GLUD1 is a 'house-keeping' gene, a general factotum operating in many tissues. It has remained unchanged since the formation of GLUD2—having accumulated no sequence changes whatsoever. GLUD2, however, has experienced a very different evolutionary history. It has been honed by evolution to operate effectively in the brain. Burki and Kaessmann identified two amino-acid substitutions in the GLUD2 protein. One of them allowed the GDH enzyme to work furiously—taking its metabolic brakes off. The same mutation also allowed GDH to be very active in the brain in the slightly more acid conditions caused by intense firing of neurons. GLUD2 has been turned into a race-horse. The more advanced the brain, the more intense the neuron signalling, or traffic, is likely to become. GLUD2 allows very rapid use and recycling of the crucial enzyme GDH—so, to mix metaphors a bit, it is like swapping a Lee-Enfield rifle for a machine gun. But it is not only the speed of nerve impulses that is involved. GLUD1 has been shown to increase its activity during periods of memory formation in rats. Perhaps this means that GLUD2 underpins some more specific aspects of enhanced hominoid cognition? Burki and Kaessmann are convinced that GLUD2 is one of a number of genes that laid the foundation for the later expansion and specialization of the human brain.

Since 2006, the idea has been growing that structural variation between genomes might be a far more widespread and important feature of evolution than previously thought. It is set to dwarf studies of both gene expression differences and single point mutations within genes as a potent explanation of the processes that make individual humans different from each other, and of what has made us human in the first place. We have discussed the idea that genes can make copies of themselves, via reverse transcription, and plop them back into the genome where they occasionally thrive. The biggest growth area in comparative genomics—some say nothing short of a revolution in genetics—has discovered two important 'more is better' mechanisms in genomes. The first is that genes can often make multiple copies of themselves—this is called copy number variation. The second is that whole sections of DNA sequence, often many thousands of DNA bases in size, can be copied and re-inserted elsewhere in the genome—this is called segmental duplication. To use the crusty old metaphor of the human genome being the 'Book of Life' we are rapidly moving from a position where we thought evolution was equivalent to changing a few letters in some of the words (single point mutations), to where some words are written in a bold type-face (changes in gene expression), to a position where whole sentences, paragraphs, and even pages of the book are repeated many times and re-inserted all over the place. The genome is rapidly becoming a much more exotic and dynamic place, and this has enormous repercussions for our understanding of the real nature of chimp–human differences.

The sheer scale of human copy number variation was unveiled in November 2006. Several scientists from the Sanger Institute in the UK, including Matthew Hurles, were involved. When the 30,000 or so genes in the human genome were compared between a number of individuals they identified 2,900—that's 10%—as having either higher or lower numbers of copies of themselves relative to the average of the group. Humans, previously thought to be about 99.95% similar, regardless of ethnic group, are now seen as being much more variable—less than 99.5% similar.

When the team looked up an authoritative online database of disease-related genes they discovered that 10% of them existed as multiple copy number variants. Genes involved in brain development and the immune system were particularly variable in copy number. Of particular interest was a gene called *CCL3L1* which, when present in multiple copies, is associated with resistance to HIV infection. Alternatively, it leads to susceptibility to HIV if copy number is reduced. *CCL3L1* controls production of a chemokine (an immune system molecule which organizes white blood cells to fight infection), which is associated with the *CCR5* receptor molecule on the cell surface. In chapter 5 we identified *CCR5* as a method of entry into cells by the HIV virus and discussed the fact that a nonsense mutation, *CCR5 del 32*, reduced susceptibility to AIDS by reducing the population of this receptor. *CCL3L1* seems to have the opposite effect—whereas 'less is more' is the motto for *del 32* (in terms of resistance to AIDS), 'more is better' could be the motto of *CCL3L1*, because its interaction with the *CCR5* receptor is antagonistic to HIV and prevents entry. A research team from the University of Texas calculated that each extra copy of *CCL3L1* decreased the risk of HIV infection by between 4.5% and 10.5%.

Copy number of *CCL3L1* varies between ethnic groups and between individuals within each ethnic group. It can vary from no copies to five or more. You are more susceptible to HIV if your copy number is lower than the average for your ethnic group. Copy number is particularly high in Africa. The Texas researchers further discovered a fascinating dual effect for *CCL3L1* and *CCR5*. They have identified what they call a 'permissive genetic background' for AIDS in individuals who lack the *del 32* mutation and have low copy number of *CCL3L1*. They have a threefold risk of rapid progression to eight of the twelve AIDS-defining illnesses such as Kaposi's sarcoma and cryptococcosis, whereas a combination of the *del 32* mutation and high relative copy number of *CCL3L1* have a more 'protective' genotype which confers resistance and slows progression of HIV to full-blown AIDS. An extraordinary modern 'combo' of the 'less is more' and the 'more is better' principles!

If the role of copy number variation in human genetic variability makes any one human less closely related to any other, what does this mean for the genetic relationship between humans and chimps? From several studies recently reported it is clear that the genetic distance between the two species is amplified. Take a large chunk of matching coding DNA sequence from each of the human and chimpanzee genomes and you will find a number of points at which a single DNA base has been substituted between the two species, corresponding to the classic 1.6% or so difference. However, if you factor in the variation due to copy number differences, chimps and humans end up not 98.4% similar but closer to 96%.

Prof Jim Sikela, of the University of Colorado Health Sciences Center, has championed research on copy number variation between humans and chimps. For Sikela, genome research is akin to a little girl discovering her mother's jewellery box for the first time, and revelling in the cascade of twinkling riches within. In another of the great landmark papers of comparative genomics, appropriately titled 'The Jewels of our Genome', Sikela has detailed the evolutionary importance of structural variation. Back in 2004, together with colleagues Andrew Fortna and Jonathan Pollack, and others, he made the first genome-wide comparison of copy number variation between humans and the other higher primates. It was a test of his belief that copy number variation has been *the* prime mover in hominoid evolution and the divergence of humans from chimpanzees.

Before we look at what Sikela, Fortna, and Pollack found, it is well worth asking the question: Why has it taken so long for copy number variation to emerge as the main contender? Sikela explains that the reason why earlier chimp–human genome comparisons suggested very close proximity between the two species in the first place is down to the limitations of genome technology until very recently. This forced earlier investigators to look at very limited regions of the genome—sometimes single genes—and this inevitably led to the bias toward looking at single base-pair mutations. We have had to wait for the arrival of a new technique called array-based comparative genomic hybridization to improve matters. In this technique the DNA of the species to be compared is tagged with fluorescent dye and hybridized to a two-dimensional array of thousands of human genes on a glass plate. The success with which the comparison DNA matches, or hybridizes with, the human DNA can be measured by intensity differences in the fluorescence of the dye. Areas of high fluorescence intensity indicate multiple copy number of the matching gene.

This revolutionary technique was used by Fortna, Pollack, and Sikela to compare 30,000 genes common to humans and the four great apes: chimps, bonobos, gorillas, and orang-utans. Each ape species' DNA was tested against human DNA as the reference: bonobo/human, chimp/human, and so on. In this way human DNA was compared with all the other primates. Out of those 30,000 genes, they identified 815 genes that showed either comparative increases or decreases in copy number that were unique—they appeared in one species and were shared with no other. There was substantial difference between the species in the ratio of genes that had increased copy number and those that had decreased copy number. These broke down as follows: human 134/6; bonobo 23/17; chimp 11/4; bonobo/chimp combined 26/11; gorilla 121/52; orang-utan 222/188. There was a general tendency for the total number of copy number variations to increase as you go back in time from the human lineage to gorilla and orang-utan, i.e. to some extent the amount of copy number variation was a function of time. However, the comparison between chimpanzees and bonobos on the one hand, and humans on the other, was exceptional and instructive. Although the ages of the human lineage and the chimpanzee/bonobo lineage are the same—at 6 million years since the split from the common ancestor—the human genome has acquired vastly more copy number variations than the chimp, and the vast proportion of them have resulted in an increase in the number of copies of genes. In humans, the ratio of copy number increases over decreases is 22:1, whereas for chimps it is less than 3:1.

The value of gene duplication to evolution is that increasing copy number of a gene can quite simply greatly increase the amount of gene product—protein—that can be produced. More interestingly, once an organism has duplicated a gene, perhaps many times, natural selection can act on each copy of that gene at different times and rates. Gradually, one or more genes in a family descended from one ancestor can differentially accumulate mutations which can change the nature of the protein it codes for, and thus the biological processes with which that protein is involved. These 'dupes' in the human genome are far more frequent than previously realized, accounting for at least 5% of the coding sequence of DNA in our genome.

Fortna, Pollack, and Sikela looked in more detail at the 134 genes showing copy number increases solely in the human line. Only half of them were genes whose biological function was known. They discovered a cluster of duplicated genes in a very unstable region of chromosome 5 including BIRC1, which is thought to delay the programmed culling of nerve cells. Could a boost in this gene lead to an increase in neuron proliferation and/or an increase in brain size in humans? Several other genes were found to be active in the brain and involved in the transmission of signals along nerves, promoting long-term memory, and the growth of the synaptic connections between nerve fibres. They also found that the *aquaporin* gene, AQP7, which is involved in water transport across membranes, shows increased copy number unique to humans. It has been duplicated several times and all those copies appear functional. This suggests significant selection pressure has been operating. Sikela speculates that aquaporin evolution helped to mobilize glycogen energy stores during long periods of sustained endurance running that turned early humans into effective savanna predators and scavengers. Aquaporin also facilitates sweating—which would have been vital to dissipate all the heat that running caused and which might also have cooled the expanding human brain and contributed to the selection for human hairlessness.

What about genes with higher copy number only in the great apes? Chimps are known to have been the biological reservoir for HIV and are immune to AIDS. So it is interesting to see that several genes associated with immune function have increased in copy number in chimps. In gorillas, high copy number maps onto habitat and diet. The gene *FLJ22004*, which is involved in cellulose and pectin digestion, has increased copy number uniquely in gorillas which chimes very nicely with the fact that they mainly eat leaves, and fruit when they can find it. Gorillas also, uniquely, have many copies of genes in a family of enzymes that are important for breaking down toxins. This is clearly important if your diet includes over one hundred plant species, many of which contain poisons.

Carrying on from this earlier work, Sikela's team further identified one gene, *MGC8902*, which showed a most striking human-specific increase in copy number. It has 212 copies in humans and half that number in chimpanzees! This gene codes for six DUF1220 protein domains. A protein domain is a complex structural element of the protein chain—it will be very closely related to how the protein interacts with other molecules. DUF stands for 'domain of unknown function'—so we have another mystery on our hands. Only primates have multiple DUF domains, these sequences show signs of positive selection, and the copy number or amplification increases through the primates as you go toward humans. However, DUF1220 domains turn up in a number of proteins, coded by a number of genes, so, not only has DUF1220 proliferated thanks to some frenzied copying of *MGC8902*, but thanks to the copy number increases of a whole suite of genes. It is clearly very important, disproportionately, to humans, and has been strongly selected for.

But what does DUF1220 do? It turns up in neurons, particularly the Purkinje neurons in the cerebellum, which are among the largest in the brain. These neurons possess a rich thatch of branching dendrites—a row of them looks like a brake of trees. They are extremely important for motor or muscular coordination. In addition to the cerebellum, DUF1220 appears in the hippocampus, that part of the brain that turns emotional experiences into memory, and it is expressed very abundantly in neurons in the frontal, parietal, occipital, and temporal lobes of the neocortex, all of which are essential to higher cognitive function.

How many of these genes with increased copy number are really responsible for what makes us human? There is clearly a growing list of genes that have uniquely increased in copy number along the human line, and many of them are expressed in different parts of the brain. Development of cognition and overall speed of neuron firing are often implied, but, as yet, far from satisfyingly nailed down. Sikela's research is like an exciting shopping list of copy number events that look suspiciously like the hand of natural selection at work, but it will require a great deal of follow-up work to find out which of these multiply copied genes have really accelerated the pace of human evolution, speeding us away from our nearest living relatives. However, in 2007, one piece of research quite literally gave us food for thought because it involved the mouth and digestion, not the brain. However, if the authors are correct, it may tell us a great deal about evolutionary change peculiar to our early human ancestors that began to fuel brain expansion. Or, as *Science Daily* put it: 'To think that world domination could have begun in the cheeks!'

George Perry is an expert on copy number variation, and Nathaniel Dominy is a professor of physical anthropology. Dominy believes that diet is the clue to the rapid and huge expansion of the human brain and he is particularly interested in the period of dramatic brain expansion around 2 million years ago, when *Homo erectus* was the prominent hominin. The human brain is a very expensive organ to grow and keep functioning, but, around this time it did grow, while the size of the gut diminished. Together, this suggests that our human ancestors must have changed their diet away from relatively indigestible material like leaves and other vegetation, toward more energy-rich food materials. But what? Until recently it has been assumed that 'something' was meat. There are as many scenarios as there are heaps of discovered animal bones in the Rift Valley, but it is quite clear that there was some evolutionary trajectory from a mainly tree-living great ape that ate mainly leaves and fruits, to *Australopithecines* with massive jaws and dentition, to *Homo habilis*, at about 2.5 million years ago, and *Homo erectus*, who were increasingly gracile and the proud owners of the first stone tools. The following scenario illustrates the general idea.

One of the great bug-bears of human origins research is that there are simply not enough fossils, neatly strung out through time, to get a continuous picture of what was going on. But there are truckloads of animal bones, particularly those of the great herbivores. By carefully analysing the tooth and bone structure of large numbers of antelopes, Elizabeth Vrba, from Yale University, has concluded that many species, adapted to chewing leaves from trees, simply went extinct about 2.5 million years ago. They were replaced with grass-adapted species—which can be deduced from the size and crown structure of their teeth. Giant buffaloes and hartebeests also arrived, together with horses and black buck. Others have documented a rapid pulse in extinctions and radiations of a number of rodent species at exactly this time. Vrba calls this the 'turnover pulse', a huge response to changing climate and landscape. It was also the pulse that produced *Homo habilis*, with his larger brain, bigger stature, and the first, Oldowan, stone tools.

With the arrival of massive grasslands, and the reductions in trees and fruits, the question became: What is the best way to eat grass? Two species of early hominin existed side by side at the time—and each went about the problem in an entirely different way. The robust *Australopithecines* developed awesome dental machinery capable of crushing the biggest nuts and tubers and pulverizing the harsh, leathery grass into an edible pulp. But *Homo habilis*, smaller and more gracile, adapted to eating grass indirectly: he let grazing animals eat the grass, then scavenged the carcasses of the animals from under the noses of hyaenas and other big carnivores. The simple flaked pebbles of the Oldowan toolkit took the dental machinery out of his mouth and put it into his hand. It meant *habilis* could now open the hardest of containers—bones—to get at the succulent, and highly nutritive, bone marrow. This latter tack proved to be decisively the more effective. Rapid climate change caused wild fluctuations in types and abundance of vegetation, but by eating animal species, in changing times, *Homo habilis* was able to buffer himself against change, stabilize his diet, and, in time, evolve so as to reduce the amount of gut he needed to digest all that plant-food. He used the metabolic energy to grow grey-matter instead.

… Well, maybe. Others have worried about just how much meat, or marrow, these early hominins could have got their teeth into, and, indeed, whether their smaller teeth would have been terribly efficient at ripping and tearing at raw carcasses—even with a little knapped flint to help. Modern human hunter-gatherers actually eat very little meat, and Dominy thinks it was ever thus:

> Even when you look at modern hunter-gatherers, meat is a relatively small fraction of their diet. They cooperate with language, use nets; they have poisoned arrows, even, and still it's not that easy to hunt meat. To think that, two to four million years ago, a small-brained, awkwardly bipedal animal could efficiently acquire meat, even by scavenging, just doesn't make a whole lot of sense.

What does make sense to Dominy is starch, in the form of tubers or bulbs, or any kind of underground plant storage organ. Field research in Tanzania by John Moores, from the University of California, San Diego, lends weight to Dominy's theory. Moores has been documenting the use by chimps of rudimentary tools like sticks and pieces of bark to dig for edible roots, tubers, and bulbs in the arid woodland savanna of the Ugalla region. This is an area many researchers believe resembles the open savannas thought to have been created in east Africa over 2 million years ago by climate change which brought dryer, cooler conditions and shrank the forest. In Ugalla today the climate is seasonal, shifting from lush, wet winters to dry, hot summers. The vegetation is like an open deciduous woodland—belts of trees surrounded by large swathes of open grassland. Today's summers re-create the likely environment our human ancestors would have encountered when, like today's savanna chimps, they may have been forced to rootle in the dry earth for their food.

For some years, Dominy has been doing field research with the Hadza, a tribe of hunter-gatherers living around Lake Eyasi in Tanzania, and has noted that they get up to 40% of their daily energy input from tubers, which they cut up and roast over open fires. To digest starch we produce copious amounts of the enzyme amylase, produced in the salivary glands. Could it be, he thought, that those peoples who subsisted on diets rich in starch would have differences in the gene, AMY1, which codes for amylase, compared to those peoples whose diets contained more protein? Specifically, had the different food ecology of different human groups given rise to differing selection pressure on the amylase gene that might be reflected in its copy number?

Although there is a great deal of dietary variation around the globe it is possible to distinguish between ethnic groups that are 'high starch' and groups that are 'low starch'. They chose to compare copy number between high-starch populations—European-Americans and Japanese, who eat cereals and rice respectively, and Hadza hunter-gatherers, who eat starch-rich tubers—and low starch populations, which subsist on meat, blood, and fish or sugars derived from fruit, honey, and dairy products. These included the Biaka and Mbuti, who are rainforest tribes; the Datog, who are Tanzanian animal-herders; and the Yakut, a Siberian tribe who subsist by fishing. They found that twice as many individuals in the high-starch groups had six or more copies of AMY1 as in the low-starch groups.

How might greater amylase production have given those populations and individuals with higher copy number and starch-rich diets a selective advantage? Amylase pours out of the salivary glands into the mouth when food is chewed. It pre-digests starch in food and a significant amount of glucose is liberated from the starch before the food even reaches the stomach. That glucose can be absorbed directly into the bloodstream. This could be particularly valuable if the individual concerned is suffering from diarrhoea. As the authors point out, 15% of child deaths under the age of five, world-wide, are caused by diarrhoea. Absorption of glucose directly into the blood from the mouth cavity could be a life-saver at a time when the contents of the gut are being evacuated at high speed. Also, it has been shown that salivary amylase persists in the gut for a long time, valuably augmenting pancreatic amylase activity in the small intestine.

Now comes the clincher. Perry and Dominy analysed amylase copy number in chimpanzees. There was no variation at all. Chimps have only the normal diploid number of AMY1 genes. Bonobos do seem to have some increase in copy number, but these gene copies appear to be non-functional. So, the average human has at least three times more AMY1 copies than chimps, while bonobos, who feed intensively on fruit, may not make amylase at all.

This study cannot yet resolve precisely how diet drove the evolution of the human brain. Was it meat and no carbo-veg? Carbo-veg and no meat? Or meat and two veg? Was it cooked, or eaten raw? Rough estimates from Perry and Dominy, based on the small amount of DNA sequence divergence between the multiple copies of the AMY1 gene, suggest a recent origin for the copy number increases, perhaps as recently as 200,000 years ago. If true, this would not be relevant to brain expansion in *Homo erectus* 2 million years ago. It would, instead, suggest that the value of flooding the mouth with amylase had to wait until the arrival of modern humans who, presumably, had fire to cook either or both meat and tubers, which made the food easier to digest and required less energy expenditure to do it. However, the authors admit that their data on the age of the copy number proliferation is simply not good enough to make reliable estimates. Data on AMY1 sequences from many more individuals will be needed to allow them to pin that date down—and this might throw the date of the copy number increases back in time such that periods of known human brain expansion, diet, and ecology knit together into a persuasive picture of how we ate our way to being human.

Matthew Hahn, Jeffrey Delmuth, and Sang-Gook Han have recently looked at both gene gain and loss in primates. You need to look at gain and loss together, they argue, because what is important is genetic turnover: humans 'churn' their genomes at a far greater rate than chimpanzees and all the other primates. They have identified several brain-related families of genes that have more than doubled their size uniquely in humans and show that, in many of these families, there is an increased level of positive selection on their nucleotide sequences. They calculate that there has been a gain of 678 genes in the human genome and a loss of 740 genes from the chimp genome since the common ancestor some 6 million years ago. This means, they say, that an astonishing 6.4% of all human genes, 1,418 out of 22,000 or so, do not have a one-to-one matching copy in chimpanzees. This 'genetic revolving door' must certainly account for unique human adaptations. The accelerated rate of evolution in primates, and particularly humans, they say, suggests that duplication and loss of genes has played at least as great a role in the evolution of modern humans as the modification of existing genes. They specifically conclude:

> These findings may help explain why humans and chimpanzees show high similarity between shared nucleotides yet great morphological and behavioural differences.

The infamous 1.6% counts for little.

Copy number variation can vastly increase the presence of one original gene throughout the genome, greatly amplifying its biological effect. But larger stretches of DNA can also get duplicated and dispersed all over the genome too. These are the sequence duplications, SDs for short, and it is becoming rapidly clear how potent a force for evolution they are, and how unstable the genome is because of them. Humans and great apes are different to other animals in that their genomes are littered with these large, interspersed segmental duplications with high levels of sequence identity. It looks as if distinct waves of duplication have taken place during primate evolution. The whole phenomenon may be pivotal, from an evolutionary stand-point, precisely because of the genomic instability and large-scale chromosomal rearrangements this massive duplication and relocation of DNA causes. Evan Eichler's group at Washington University, and others, have looked extensively to see where these segmental duplications occur in the genome, and have found that they do not occur at random—there are duplication 'hot-spots'. Eichler's group call them 'duplication hubs' or 'duplicons'. In humans and great apes around 450 duplicons have been identified which have recruited duplications from a number of ancestral loci elsewhere in the genome such that these hubs have become, in Eichler's words, complex mosaics of different genomic segments where novel genes, fusion genes, and gene families have emerged.

These hot-spots are associated with rapid changes in mutation rate among genes, major increases in gene expression, and a whole range of genetic diseases. They are literally crucibles for rapid evolutionary change. This is the signature of evolution-in-a-hurry. When it comes to evolution, particularly and peculiarly in humans, the motto could well be that of the slap-dash chef who blithely says, 'Well, you can't make an omelette without breaking a few eggs!' Genetic diseases are the broken eggs in the kitchen that is the human genome. They are the price humanity pays for the rapid pace of evolution in the human line since the split from the common ancestor.

Eichler and his group had first examined the distribution throughout the genome of sequence duplications that were nearly identical, between 90% and 98% similar, and greater than 10,000 DNA bases in length. They found that some 5% of the genome consisted of these large duplicated segments, often containing copies or part copies of genes, as opposed to genetic rubbish. The size, fraction of the genome, and sequence identity of these dupes caught most scientists by surprise. Most believed any duplication would be restricted to occasional clusters of genes called tandem repeats or to oddball areas of the genome like the Y chromosome. But Eichler and others found these large duplications were spread throughout the chromosomes. It also became clear why these large, near identical, segmental duplications had important repercussions for disease. During meiosis, the process of cell division that produces gametes, there is an initial period called prophase where the two chromosomes of each pair briefly align together. This allows 'crossing over'—genetic material is exchanged between the chromosomes. Normally, the identical stretches of sequence from both chromosomes find each other so this initial alignment occurs with great accuracy. However, the very high degree of sequence identity in duplicated segments can cause confusion. Sometimes the chromosomes misalign by confusing one matching stretch of DNA for its near-identical dupe. These misalignments can lead to either the deletion, duplication, or inversion of DNA sequence and cause any number of genetic diseases like Williams-Beuren Syndrome and Angelman Syndrome. Other examples of these genetic disorders include DiGeorge syndrome, one of a range of syndromes known as CATCH 22 because they result from a deletion of a tiny part of chromosome 22 known as 22q11.2 which causes several genes to be lost. DiGeorge syndrome includes abnormalities of the immune system and parathyroid glands and congenital heart defects. But don't forget the upside. These structural upheavals can also have the constructive effect of creating entirely new genes.

Evan Eichler is a pioneer in this area. He and his team have discovered a beautiful example of the upside of segmental duplication, which has produced a whole family of novel genes out of what appear to be functionless areas of the genome. The new gene family has arisen because a 20,000-base segment of DNA has multiply copied itself and spread those copies throughout a much larger 15-million-base area of the short, or 'p', arm of chromosome 16. There have been flurries of duplication before and after chimpanzees and humans diverged from each other which have resulted in a family of 17 copies of the duplication in gorillas, between 25 and 30 duplications in chimps, and 15 duplications in humans. Within these blocks of duplicated sequence an ancestral gene had evolved into a family of new genes. In a flight of literary fancy unusual among genomics researchers they named the gene family *Morpheus*, after the god in Ovid's '*Metamorphoses*', who sends dreams in human forms. Comparison of the exon sequence, the coding part of the gene, revealed that evolution must have happened at a gallop, causing major episodes of amino-acid substitution before and after the divergence of humans and chimps 6 million years ago. Amino-acid changes in the resulting proteins accumulated at fifty to a hundred times the normal mutational rate! Surprisingly, there had been much more mutation in the coding exons (10%) compared with less than 2% in the non-coding introns—it is usually the other way round. This was the signal of wildfire evolution. The most dramatic difference was between exon 2 for humans and chimpanzees compared to Old World monkeys. The level of amino-acid replacement translated into a whopping 43% amino-acid divergence, 20 times what you would expect. Humans then accrued a further 23% of amino-acid divergence compared to chimpanzees!

The *Morpheus* gene family represents an evolutionary leap between Old World monkeys and great apes, and further to humans, of almost preposterous proportions. Although the researchers, to date, have very little specific idea of what the *Morpheus* gene family actually does, it seems likely that the ancestral gene was restricted to some function in the testis and that, over 25 million years of being hurled around the chromosome in a succession of duplications, sequence shufflings, and other rearrangements, family members have acquired novel functions and novel regulatory machinery. They may very well play some fundamental role in the immune system.

Most of the changes in the genetic landscape between chimps and humans, according to Eichler, can be attributed to duplications. They have altered some 2.7% of the genome compared with the 1.2% genome alteration caused by single base-pair mutations—so they are over twice as important as 'snips' in causing genetic change. The frustrating thing about all this research is that no-one yet has the full story—merely the opening chapter. There is still a dearth of examples of where structural duplication has led to a novel gene which has underpinned some novel, evolutionary function and created species differences. It is a case of 'watch this space'!

It is beginning to look as if the evolution of the immune system is driven by the sort of gene duplication we have seen associated with the *Morpheus* gene family. Tatsuya Anzai, and a host of Japanese collaborators, have focused on the MHC—the Major Histocompatibility Complex—a stretch of over 220 genes on chromosome 6. The MHC is a vital part of the body's immunological defence system and is involved in the ability of cells to distinguish between 'self' and 'non-self' and therefore mobilize antibody defences against foreign antigens. Many of these genes are highly polymorphic, which means there are many variants of them. Some of them have over 400 alleles and this extraordinary level of diversity is necessary to allow the immune system to challenge the immense range of pathogenic antigens our bodies are constantly bombarded with. Anzai looked at part of the Complex which is known to contain several multicopy gene families that have come about through repeated bouts of gene duplication.

The team sequenced 35 genes from this area that are shared by humans and chimpanzees and found DNA nucleotide similarity to be 98.9% and amino-acid similarity in the resulting protein to be 98.3%. However, 28 of those genes, although present in the Major Histocompatibility Complex, had no immune function at all—they did something else. When the team separated them out from the remaining immune-system genes, they found that amino-acid sequence identity was over 99% for the non-MHC genes but only 95% for the seven immune-system genes. This means that evolution has operated to maintain sequence similarity between human and chimp with respect to these non-MHC genes but has operated to promote sequence diversity in the immune-system genes—which makes sense if their job is to respond to a great range of infections caused by different microbes in chimpanzees and humans.

These differences between the two classes of gene were amplified still further because the genes are differentially peppered with little inserted elements of 'junk DNA' called ALU repeats, LINE elements, and SVAs. When they took these differences into account the sequence similarity, which had been around 98.6%, fell dramatically to 86.7%. A 12% difference between the two species had been caused by these inserted elements alone! If this much lower figure of 86.7% sequence identity were to generalize across the whole genome it would re-write popular accounts of chimp–human differences as 'the 13.3% that makes us human'!

A recent genome-wide comparison of segmental duplication between human and chimpanzee has calculated that one third of all human sequence duplications do not occur in the chimpanzee. It found 32 million bases of duplication specific to humans but 36 million bases of duplication specific to chimps. This had duplicated 177 genes in humans, but not in chimps, and 94 genes had duplicated in chimps alone. Over half the human-only duplicated genes had up-regulated, or increased, their gene expression, whereas a slightly lower percentage of chimp-only dupes had up-regulated. It is clear that the genomes of the two species have been accelerating away from each other since they diverged 6 million years ago. At least 70 million bases of actively coding DNA have been differentially duplicated in chimps and humans, and this is equivalent to 2.7% of the genome.

Jim Sikela lists 28 genes that show copy number changes specific to the human line that have come about due to segmental duplications. Most of them are involved in the brain and central nervous system. This is almost certainly only a tiny fraction of brain-expressed genes that have evolved specifically in humans due to structural rearrangements of our genome. Neuroscience has not even scratched the surface. Sikela quotes Thomas Insel, director of the US National Institute of Mental Health, as saying that 99% of the current neuroscience literature focuses on just 1% of the genes expressed in the brain! And we have not yet finished exploring the range of structural differences between chimpanzee and human genomes. Jim Sikela's 'jewellery box' is turning into Aladdin's Cave!

CHAPTER 7

Aladdin's Cave

In the previous chapter we saw that gene duplication and segmental duplication—even by the most conservative estimate—each contribute twice as much to the genetic divergence of human from chimp as single point mutations. We also saw that these duplication hubs are hot-spots for genome evolution and that genes within them exhibit much higher levels of mutational change and gene-expression changes than genes elsewhere. It also transpires that the answer to the question 'how similar are the human and chimpanzee genomes?' depends on where you look, and that the sequence divergence in many of these duplication hubs is so great that, in these areas at least, the similarity between the two genomes is driven down from the old estimates of 98.4% (or thereabouts) to as little as 87%. And we are not yet finished with structural genome and gene variation. Over the past decade, armed with huge, powerful new tools of computational investigation, genome scientists have discovered more and more 'jewels in our genome' and the potent role they play. To carry further Jim Sikela's metaphor that genome investigation is akin to a child's exploration of a jewellery collection, the genome is rapidly turning out to be a veritable Aladdin's Cave.

Two of the by-products of the structural tectonics played out in chromosomes are insertion and deletion. Deletion is where a whole chunk of DNA code has simply been dropped from the sequence; insertion is the opposite. Insertions and deletions are known collectively in the trade as 'indels'. These are important because insertion or loss of DNA from a gene will cause a frame-shift mutation, where the code, in effect, jumps, or where the new inserted sequence results in a punctuation mark where transcription into the messenger RNA comes to a premature halt. In 2003, a research team from Perlegen Sciences, headed by Kelly Frazer, compared the occurrence of indels on human chromosome 21 and its chimpanzee equivalent—chromosome 22. They identified 57 indel differences, ranging in size from two hundred to eight thousand DNA bases, between the two species. They concluded, from the 27 million bases of chromosome 21 they had covered, that indels add a further 0.6% of sequence divergence between humans and chimps to the 1.2%+ for which single nucleotide changes are responsible—swelling genetic divergence by 50%. It looks, from their data, as if species divergence has been a two-sided affair—both species have moved away from each other. They looked at sixteen indels in greater detail, using orang-utan DNA sequence for comparison. (If the chimp and orang-utan DNA matched, the indel must be unique to humans, if the human and orang-utan DNA matched, the indel must have arisen in the chimp.) They discovered that three insertions and two deletions had arisen in the human line and ten deletions and one insertion had occurred exclusively in chimps.

According to Jim Sikela, if you look across the genome, you find around 5 million small to moderate-sized indels in each species. The consortium of scientists charged with the first reads of the chimpanzee genome, he reports, discovered that each of those parts of the genetically active, regularly transcribed chimp and human genomes—called euchromatic sequence—contained about 45 million non-shared, species-specific DNA nucleotide bases, which corresponds, he says, to indel differences equivalent to 3% of the total genome in each species. This eclipses Frazer's estimate because it would be twice the contribution to human–chimp differences of single point mutations.

Then there are inversions. These occur at a range of scales from gross chromosomal inversions, visible down the microscope on specially stained slides, where a portion of a chromosome flips end on end, down to flips of one hundred thousand DNA base-pairs or less. In 2005, a group led by Stephen Scherer, of the University of Toronto, using computational techniques, greatly added to the list of known inversions and raised the probability that yet another source of structural change in the genome might provide fodder for evolution and be a cause of genome difference between chimpanzees and humans. Small inversions have been notoriously difficult to spot, but by computationally aligning the draft assemblies of the two genomes, says Scherer, they can not only see the big inversions but resolve differences right down to the break-point itself. To the known list of nine inversions, big enough to be seen down the microscope, Scherer's computational comparison of the two species' genomes turned up a further 1,576 probable inversions. Thirty-three of these were larger than 100,000 base-pairs in size and twenty-nine of them intersected genes. If a break-point of the inversion occurs inside a gene, it will clearly disrupt its function. Genes adjacent to the break-points may also be affected. This is why Scherer became interested—such disruption can lead to genetic diseases. On the other hand, this re-structuring may be yet another way of rearranging genetic code to provide the raw material for natural selection to work on. Again, this is the two-edged sword of human genomic evolution. Twenty-three of the inversions identified in a computational sense have been confirmed experimentally, with the largest being 4.3 million base-pairs in size on human chromosome 7. Surprisingly, three of the inverted regions were found to be variable in their orientation in the human population. For instance, some of the people tested displayed two alleles (paired variants of the same gene) containing the human-specific inversion, whereas others were heterozygous—they had one allele containing the inverted sequence and the other allele containing the ancestral non-inverted sequence, identical to chimpanzees. They discovered that the heterozygous variant had, in each case, been inherited by the individual's mother and that the inversion region covered more than fifteen genes of which one, PMS2, is known to be involved in colorectal cancer.

Altogether, over half of the 23 experimentally validated inversions were found to be unique to humans. They also established a high correlation between inversions and segmental duplications. We have already established segmental duplications as hot-beds of genome re-assembly and it seems as if inversions join indels and copy number increases in the family of structural changes we are beginning to associate with these areas. The link with genetic disease is that the orientation of inversions causes the chromosome alignment problem mentioned in the previous chapter: if one sequence is inverted and its sister sequence on the homologous chromosome is not, the two chromosomes may not be able to align themselves properly during meiosis and this may lead to important and potentially catastrophic deletions of DNA code—causing genetic disease. Scherer had done earlier work on the genetic disease Williams-Beuren Syndrome, sufferers of which are mentally retarded. They have distinctive, and quite attractive, pixie faces and, despite their very low IQs, extraordinary powers of empathy with people and fluent, idiosyncratic use of language. It is in many ways the opposite of autism—sufferers find people as interesting as some people with autism find railway timetables. Although Williams-Beuren Syndrome is caused by a micro-deletion of DNA sequence, the parents of Williams children show an increased incidence of inversions. The same applies to Angelman Syndrome and Sotos Syndrome—a form of gigantism characterized by excessive growth in the early years and some degree of mental retardation. The fact that Scherer and his colleagues have discovered these inversion variants suggests the human genome is still in flux and that, in this sense, it is still evolving. There have clearly been a great number of these inversion events since we parted company with chimpanzees, though we wait to see if any strong candidates for adaptive gene evolution, the other side of the coin to genetic disease, emerge.

Inversions may have a yet more fundamental and important role in primate and hominid evolution. They may be the mechanism that caused the split of chimpanzee ancestors and human ancestors from a common ancestor. Such speciation can occur where individuals of a population become physically separated from the others and cannot interbreed with them because of some physical barrier— a deep river, a chasm, or a mountain range would do the trick. Gradually, over time, the two separated sub-populations will accrue mutations at different rates until they reach the point where their gametes, should they eventually meet, will no longer be compatible. Another form of speciation depends on barriers to effective repro- duction among individuals of a species that have not been physically isolated.

In 2003, Alec McAndrew considered how speciation could have occurred among a population of our common ancestor such that chimp and human ancestors became reproductively isolated. One major suggestion was that a major chromosomal rearrangement— an indel or an inversion—may have occurred. Certainly, if you look at human chromosomes 1, 4, 5, 9, 12, 15, 16, 17, and 18, all have large inversions compared with their equivalent chromosomes in chimps, and human chromosome 2 is a fusion of two chimp chromosomes. Individuals who bore one original chromosome and one rearranged chromosome—called heterozygotes—would be far less fertile than homozygotes who bore two identical chromo- somes because recombination between the rearranged chromo- some of a pair, and the unaltered one, could result in catastrophic misalignment. Selection would thus favour the homozygotes and two populations, rearranged and not, would emerge and be reproductively isolated.

However, McAndrew spotted a flaw in this argument in that, if heterozygotes are much less fertile than normal homozygotes, in which the rearrangement has not occurred, it is hard to see how the rearrangement mutation can avoid being rapidly selected out. If they are not less fertile then the rearrangement mutation would happily become established and not be able to act as a breeding barrier and the resulting population would be made up of individuals bearing either rearranged or original chromosomes.

The problem is solved, says McAndrew, if you assume that the rearranged chromosome simply does not recombine with its unaltered sister chromosome during meiosis. So, for the first X generations, offspring would be viable even though one pair of chromosomes never recombined to exchange material and sat on the side-lines. However, this effectively isolates the rearranged chromosome which could then accumulate mutations gradually over time in its own quiet little way until a point is reached at which these accumulated mutations had an effect on sexual compatibility. Two populations might then genetically diverge in the same geographical spot.

There is already evidence that this is what has actually happened. Arcadi Navarro and Nick Barton have argued that mutations should accumulate in chromosomes that had undergone fixed structural changes like inversions. Specifically, they showed that in rearranged chromosomes—those that had accumulated inverted sequence—the ratio of non-synonymous to synonymous mutations, the signature of positive selection, is far greater than among chromosomes that had not experienced any gross structural changes. So it could well be that, around 6 million years ago, one of the inversion-heavy chromosomes identified by Stephen Scherer and his colleagues caused the individuals carrying it to become reproductively isolated.

Those individuals—if they ever really existed—may have been our ancestors.

This story has another twist that takes us to the heart of the genome changes that make us human. In 2004, Tomas Marques-Bonet, Mario Caceres, Todd Preuss, and Arcadi Navarro banded together to look for a link between chromosomal rearrangements like inversions, and differences in gene expression and regulation. They discovered that genes on rearranged chromosomes showed much greater gene expression increases than genes on unaltered chromosomes, *but only in the brain*. This effect of chromosomal rearrangement, they say, is on a much larger scale than differences in nucleotide sequences or gene expression itself, when you compare the chimpanzee to the human genome.

Let us leave gross structural change in the genome for a moment and wander a little further into Aladdin's Cave, to find yet another novel way that chimps and humans have diverged over the last 6 million years. This involves genes that have dramatically evolved uniquely in humans, yet do not code for any protein! It is probably true to say that, to date, something like 99% of genomics research has concentrated on that tiny percentage of the genome—between 1.5% and 3%, depending on whom you ask—which contains functioning genes that code for a particular protein. But what about the other 97% or so of the genome? Is it irrelevant? Until recently the answer would have been 'probably' or 'we don't know' or 'it's just genetic junk', but these vast tracts of genetic 'wasteland' are turning out to be anything but junk DNA—already many regulatory elements have been found in non-coding regions, often exerting their effect at a distance. Some of it, at least, has a job. Now, thanks to the enormous power of computational genomics, Katherine Pollard, Sofia Salama, and their colleagues, connected to David Haussler's laboratory at the Howard Hughes Medical Institute at the University of California, Santa Cruz, have gone fishing for—and finding—functional islands of DNA sequence amid all this non-coding jumble. They have discovered a new, exciting, and important source of chimp–human genomic difference by taking the unconventional route of looking, not where protein-coding genes are, but where they are absent.

The best method of identifying functional sequence is to compare vast tracts of DNA between several unrelated species. If you find areas of sequence that are virtually identical, even though the species they reside in have been separated in evolutionary terms by tens or hundreds of millions of years, it is a sure sign that the sequence has some important, often fundamental, biological job to do. It has been conserved over time by negative, or purifying, selection, which has maintained its sequence integrity—drastically weeding out any mutations that might change its function. Pollard and her team searched the chimpanzee genome for chunks of sequence where at least 96% sequence identity had been preserved over approximately 100 base-pairs of DNA, when compared with the corresponding areas of both the mouse and rat genomes. This gave them a huge list of 35,000 regions averaging 140 base-pairs that had, in effect, been conserved throughout all mammals. By comparing these regions with the corresponding sequence in humans, they were able to see if humans followed the widespread mammalian trend of close identity, or whether any of the regions had broken rank and accumulated substitutions in the genetic code.

Pollard's team discovered 49 regions in humans where a significant amount of mutation had occurred. Although nearly all these regions were non-coding they often lay next door to known transcription factors—genes that regulate orchestras of protein-coding genes. Also, a quarter of these regions lay next to genes that are involved in development of the brain and nervous system, leading to the suspicion that these highly accelerated regions (or HARs for short) might have important effects on brain development.

One region, named HAR1, stood out. Out of 118 DNA bases, only two had been substituted in the 300 million years between chimp and chicken, whereas in the past 6 million years, since the split from the common ancestor, it has accumulated a massive 18 substitutions—a dramatic acceleration. The HAR1 region is not variable in modern human populations—all 18 substitutions have become fixed, which means everyone has them. This suggests that these substitutions have all occurred more than one million years ago. HAR1 occurs very close to the tip of chromosome 20 and overlaps two genes—*HAR1F* and *HAR1R*—which do not produce any protein but, instead, are translated into two very stable forms of RNA.

There is more than a fair dollop of scientific serendipity in this research. For a start, a search through the vast tracts of non-coding sequence in the human genome was daunting. They had little right to expect to find regions of non-coding DNA that had suddenly changed their spots in the hominid line from being very conserved to being evolutionary runaways. According to David Haussler, the online genome browser at UC Santa Cruz also played a role in that Katie Pollard was trawling through the region of the HAR1 sequence only to notice that a genome-wide bio-informatics scan by another post-doc in the same lab, Jakob Pedersen, had predicted a structural RNA gene at that location. Then, a visiting Belgian scientist, Pierre Vanderhaegen, possessed samples of human embryonic brain tissue which he agreed to use to determine where and when the *HAR1F* gene was active. It was first detected between seven and nine weeks in the part of the embryonic brain that develops into the cerebral cortex. It was particularly active in special brain cells called Cajal-Retzius neurons, which appear early in embryonic development and play a fundamental role in the construction of the many-layered structure of the cerebral cortex. As we discussed in chapter 3, in the context of the probable role of the genes *ASPM* and *Microcephalin*, neurons that eventually fill the cortex are born from precursor cells and then actively migrate upwards through the layers of the developing cortex until they find their destination. During the time they are active, these Cajal-Retzius neurons seem to take up position in the outer zone of the cortex and produce a protein, reelin, which they secrete downwards forming a concentration gradient. It may be that reelin acts as a 'stop' signal to growing neurons, telling them when they have reached their allotted destination in one of the cortical layers.

Pollard and her colleagues have established that *HAR1F* is active at the same time that reelin is being produced, but they have not yet demonstrated the exact nature of the relationship. But let us assume, for the moment, that the expression of *HAR1F* really does affect the expression of reelin. How important could that be? Reelin was given its name after the peculiar reeling gait of mice which were shown to suffer from a range of structural brain anomalies, especially in the cortex and hippocampus. The normal cortex develops in a very complex way. As explained by Pasko Rakic, the world expert on brain development, the first neurons destined to settle in the human cortex are produced during the first half of gestation, deep within the brain. Shortly after their last cell divisions, these neurons migrate upwards and outwards toward the surface of the cortex where they form a sheet of cells called the cortical plate. Each successive generation of migrating neurons passes through the previous generation of migrant cells before arriving at their final destination—in the manner of a complex marching manoeuvre of military bandsmen on a parade ground where serried ranks pass through the rank ahead. In this way the cortex develops inside-out. The earliest born neurons are found in the deepest cortical layers, while the later born neurons move to the more superficial layers.

Both reelin and *HAR1F* are active in the Cajal-Retzius neurons until at least 17 to 19 weeks of gestation, at which point most of the itinerant population of neurons have finished this migration march and taken their places, producing the typical laminar structure of the cortex and shooting out their synaptic processes to create complex neuronal networks. Pollard also found that HAR1F is expressed in adult brains, in the frontal cortex and hippocampus. The latter is interesting because the production of new neurons persists here throughout adulthood, perhaps because the continual formation and retention of new memory needs the constant re-formation of networks of neurons throughout life.

The best guess at the moment is that the RNA that *HAR1F* produces acts as a regulator for reelin production. Yet again, the rule in the brain seems to mitigate against mutations that change the functions of genes, but, instead, to set in place novel ways to regulate the activity of these genes. *HAR1F* appears to be yet another weapon in the brain's gene-regulatory arsenal. Given the, as yet admittedly tenuous, links between *HAR1F*, reelin production, and the extremely complex way in which the neocortex develops, Pollard's research looks like a compelling potential example of genomic developments that are unique to humans and appear to have driven the dramatic expansion of the human brain. Her group are now working through the other 48 HARs to see what other jewels they can find.

During the process of meiosis, by which gametes are made, matching pairs of chromosomes, one from the mother and one from the father, exchange genetic material between each other. This is called recombination. It ensures that the offspring inherit a novel hybrid combination of genes from both mother and father thanks to this shuffling of genetic material. The sites where this recombination happens do not occur at random but are localized to what are called 'recombination hot-spots'. Until recently, it had been assumed that these hot-spots were very similar in both humans and chimps but an international team of geneticists in Boston and Cambridge, UK, have recently discovered that this is not so. Hot-spots are rarely, if ever, found at the same locations on chimp and human matching chromosomes. They identified 18 hot-spots in humans and 3 in the chimp—and none of them overlap. This means that the very method that creates genetic variation in successive generations of offspring differs markedly between the two species.

Anybody with a biological education has been brought up on the classic description of how a gene makes protein: DNA is transcribed

into messenger RNA, each three-nucleotide codon of which codes
for one amino-acid of the resulting protein chain. Ergo: one gene =
one protein. Life, however, is far more complicated than that. It has
been discovered that one gene can produce a whole range of proteins
through a process called alternative splicing. The different forms are
called splice variants and it is becoming clear that they are a major
new source of novelty and variability in genomes in general. If one
gene can make many proteins then the protein production of the
genome is immediately amplified—which helps to explain why a
perplexingly small number of coding genes in the human genome—
about 22,000—seem able to produce the vast range of proteins we
see. Somewhere between 50% and 70% of all human genes exhibit
alternative splicing. It allows one gene to make different proteins
depending on whether it is operating in the brain or the liver, for
instance, and it allows one gene to switch to producing a different
protein when prompted by changing environmental conditions. The
Howard Hughes Medical Institute Bulletin, in 2005, for example, reported
the research of one of its scientists—Robert Darnell—in finding that
a protein called Nova controlled the splicing of 49 different messenger RNAs from one gene in the brain to produce a huge family of
related proteins that transmit signals across the synapses between
neurons.

This major source of protein diversity means that splice variants
have also played a major role in the divergence of humans from
chimpanzees. Genes are composed of interspersed chunks of DNA
sequence called introns and exons. The exons are the coding part
of the gene—the part that is transcribed into messenger RNA and
thence to protein. Introns are non-coding parts of the gene that
are not transcribed, but that usually contain elements that regulate
transcription in the rest of the gene. When a gene is active, all the
DNA is first transcribed into RNA. This RNA is then inspected by
a structure called the spliceosome which, in effect, edits out the
sequence that corresponds to the introns by snipping it out of the

RNA transcript at specific splice sites, leaving only the exon-derived sequences. The spliceosome sometimes edits out whole exons when the protein-coding part of the gene is assembled as RNA, in a process that is capable of producing a whole family of closely-related proteins with subtle differences in the way they work in the body and the brain.

Neuropsin is a protein that is involved in learning and memory and is present in the prefrontal cortex of the brain. It exists in two forms, neuropsin I and a splice variant, neuropsin II, which contains an extra 45 amino-acids, caused by alternative splicing in exon 3 of the gene. It was already known that this longer form of neuropsin, while present in humans, was not expressed in the brains of lower primates and Old World monkeys, but recent work by Dr Bing Su and his colleagues from the Chinese Academy of Sciences in Kunming, has looked at neuropsin variation much more specifically, with interesting results. They discovered that neuropsin II is also absent in the cortex of chimpanzees and orang-utans, meaning that it has to be specific to the human lineage and must have arisen within the past 6 million years. They discovered that a single mutation led to the new splicing site which created neuropsin II and further discovered that it was pre-dated by a weakening effect on a different neuropsin I splicing site, common only to chimps and humans, which had paved the way for the neuropsin II variant to occur uniquely in humans. The authors have no clear suggestion yet as to why neuropsin II has proved adaptive in the evolution of human cognition, but they note that it is restricted to the prefrontal cortex and to certain stages in brain development, suggesting that the creation of novel splice variants in the brain has been at least as important as the human-specific accumulation of DNA nucleotide substitutions in genes like *ASPM* and *Microcephalin*.

According to Benjamin Blencowe, Mario Caceres, Todd Preuss, John Calarco, and colleagues, the Chinese scientists are correct. They have taken a global genome-wide look at alternative splicing differences between chimpanzees and humans and found that between 6% and 8% of shared genes show pronounced splicing-level differences between the two species. They note that there is little species overlap between those genes that show alternative-splicing differences and those which don't, suggesting rapid evolution of distinct sub-sets of genes in humans as opposed to chimpanzees. Many genes showing such alternative-splicing differences between the two species are implicated in control of gene expression, the transfer of nerve signals between neurons, cell death, immune defence, and susceptibility to disease. For instance, one gene, $GSTO2$, which is involved in apoptosis—programmed cell death—exists in a form in human cells which is less active, compared to chimps.

The prevalence of alternative-splicing differences between chimps and humans, they say, admits splice variation to the ever-expanding club of genomic mechanisms, ranging from single nucleotide mutations, through gene expression and regulation, to segmental duplications and copy number variation and deletion, that seem responsible for major differences between the two species. And, they say, although it has been known for some time that less than 20% of alternative-splicing events have remained identical over the 90 million years that separate man and mouse, only now has anyone thought that this was likely to be reflected in important variants that have contributed to human–chimpanzee divergence. Although some 4% of genes shared between chimps and humans show much higher rates of transcription into protein (up-regulation) in humans, these are not the genes that show alternative splicing. This suggests, these scientists say, that splice variants have arisen recently as an additional source of difference between the genomes of humans and chimps.

The ultra-modern and very powerful techniques of DNA microarrays and computerized comparative genomics that allow this study, as so many more, to be conducted, are very good at sweeping rapidly through whole genomes to give us estimates of the relative importance of, say, splice variants versus gene regulation, in driving a genetic wedge between humans and chimpanzees. And they have expanded the list of genomic mechanisms that underlie chimp–human differences to the point where we realize that single nucleotide substitutions—point mutations in the genetic code—have only a limited, and perhaps minor, role to play in the divergence of the two species and in estimates of their genetic relatedness. However, such studies lack specifics. If such-and-such a gene exists in a number of functional splice variants, does it result in important changes in the structure and function of the two species, and if it does, what exactly are they? All too often this spade-work remains to be done. The example of the *GSTO2* gene mentioned earlier is a case in point. Humans have a version in which alternative splicing has skipped exon 4 entirely. The human omission does not seem to have irreparably damaged or changed the complex folded so-called tertiary structure of the resulting protein, yet the human variant produces very little of it compared to a chimp. What might be the adaptive value of that to humans?

GSTO2 is involved in neutralizing toxic compounds in the body and in protection against 'oxidative stress' whereby the body produces excess reactive oxygen in the form of free radicals and peroxides that can cause cells to literally fall apart at the seams. In humans we associate oxidative stress with diseases like atherosclerosis, Parkinson's, and Alzheimer's, while, on the upside, reactive oxygen is used by the immune system to blow pathogens to bits. Blencowe and his colleagues point out that there are major differences in the way humans and chimpanzees age with respect to the degeneration of their brains and nervous systems. In normal humans, for instance, brain aging is associated with the production of helical filaments and neurofibrillary tangles that are greatly exaggerated in Alzheimer's disease, and these do not occur in any other primate. So, on the one hand, humans live longer than chimpanzees, but when their brains age they seem to do so in a particular and often disastrous way. Against that background we find, in humans, an under-performing *GSTO2* gene. Could it be that the bodies of young humans tolerate higher levels of reactive oxygen because they rely more heavily on it than chimps do in the fight against disease? If so, this could be another example of Nature's quick and dirty fixes where the price for an efficient pathogen-killing machine early on in life is paid for in dementia later on. For the moment we are left to speculate.

During evolution in the primates, and particularly from something that was very like a chimp to us modern humans, the key word is *amplification*. We see this time and time again in the variety of genomic mechanisms we have discussed that have brought about change and difference between chimpanzee and human genomes. Changes in the DNA sequence of master controller genes producing transcription factors amplifies difference because such genes exert their regulatory effect over a whole orchestra of subservient genes— one gene affects many. The up-regulation of genes causes them to produce more protein—it amplifies their productivity. The increase in copy number of a gene does the same thing, but it also amplifies gene differences by creating a family of genes upon which natural selection can act in a differential way. Segmental duplication creates evolutionary hot-spots in the genome where the regulation and mutation rates of genes positively soar compared to more stable and less dynamic areas. Finally, splice variants amplify the activity of genes by allowing them to produce a whole range of proteins, not just one, by the selective omission of different chunks of genetic code. All of this amplifies the differences in body, brain, and behaviour between two closely related species, not only by changing the timing and productivity of similar or identical genes, but also by creating much more delicate and complex control systems, as in the creation of the 49-strong Nova-controlled gene family mentioned earlier, which allows for incredible definition in the fine-tuning of nerve-signal processing and transmission in the brain. These amplification phenomena help to explain why two organisms with two near-identical genomes at the level of much of their DNA sequence can experience different rates of evolution, and differences in their ontogeny—the way their bodies and brains develop with age to maturity. This is part of the answer to how two such closely-related organisms, in genetic terms, can unravel into two such different animals.

Nor, I think, have we fully plumbed the depth of Aladdin's Cave. The genome is weirder than we thought and every year we find stranger and stranger mechanisms that greatly amplify the very simple picture of evolution we started with. In our exploration of Aladdin's Cave we have moved from the simplistic level of under-standing, first reported in the 1960s, that, because chimps and humans appear to share the vast majority of their DNA sequence, they *must* be very similar, both genetically and in terms of their behav-iour, to an acceptance that chimpanzee and human genomes are not similar in many important ways. A host of important genomic phenomena eclipse the role of single point mutations to underpin those species differences. Along the way, perhaps we have also found a reason why the two species' ancestors split apart in the first place, all those 6 million years ago.

A consistent theme is also emerging in terms of the genes affected by these changes. They are disproportionately involved in three key areas: brain development, longevity, structure, and function; the fight against disease; and nutrition and metabolism. Over the past 6 million years our human ancestors have developed bigger brains armed with different and more powerful cognitive mechanisms, have adapted to diets very different to those of chimpanzees, and have waged war on a host of different diseases and parasites, and in different ways. The genome is like a huge haystack in which lie dispersed a large number of needles. Every year our needle-finding technologies get more powerful and the bases for our early assump-tions of chimpanzee–human similarity more frail. Given what the last five years of genomic research has brought us it is astonishing, to me, to find respectable scientists still peddling the 'chimps are us' message on the basis of a forty-year-old concept of human genetic chimp proximity. This concept of genetic proximity has, in many quarters, led remorselessly to the argument that, since chimps are genetically nearly identical to us humans, we should logically find that they think and feel the same way we do. It is a version of a common biological fallacy known as the argument from analogy and I want to take a very hard look at it in the following two chapters.

CHAPTER 8

Povinelli's Gauntlet

I n his book *Our Inner Ape*, the primatologist Frans de Waal takes a close look at the best and the worst of human nature and invites us to consider what light our two closest primate relatives, the chimpanzee and the bonobo, can shed on it. He took the title for his book from an interview quote given by the English actress Helena Bonham-Carter, who played Ari in *Planet of the Apes*. She and the other actors, she said, had gone to a simian academy where they had learned ape postures and movements and she 'had simply got in touch with her inner ape'.

De Waal reminds us that the two apes are very different with regard to their social behaviour. Chimps are from Mars, bonobos are from Venus. Chimps live in demonic societies where long periods of calm and harmony can be shattered by blood-curdling male behaviour, where females get stomped and raped, infants killed and eaten, neighbours beaten to death, and monkeys hunted by blood-thirsty gangs armed with teeth, clubs, and even, occasionally, sharp-tipped spears. Bonobos are dominated by female hierarchies, murder and mayhem is much reduced, sex is used as a social lubricant, and fun and play are the order of the day. Do we humans act as if we were a hybrid between these two apes? asks de Waal. We have the fortune, he says, of having not one, but two inner apes, which together allow us to construct an image of ourselves that is considerably more complex than much that has come out of primatology these last 25 years. We have both the chimpanzee side in us, which precludes friendly relations between groups, and the bonobo side, which permits sexual mingling and friendship across the border-line—Dr Jekyll and Mr Hyde.

This idea reminds me of the homunculus theory to explain human behaviour. Behaviour is actually controlled, the theory goes, by some kind of 'little man', some unseen and unknowable prime mover or system 'manning the bridge' in the brain and directing actions from within. De Waal seems to have populated his bridge with two captains instead of one—two *panunculi*. Like Captain Bligh and Fletcher Christian on the bridge of the *Bounty*, or Luke Skywalker and Darth Vader battling it out with light sabres, our chimp-controller and our bonobo-controller are locked in mortal combat—Jedi Knights or the Dark Side! Will it be 'Let's get drunk, smash the place up, rape a few girls, and take on the Peel Dem Crew just for good measure!', or 'Let's chill out, take in a few sounds, turn the lights down low, and make love all night long!'

De Waal asks: Are we more like chimpanzees or bonobos, which is the best ancestral model? And the answer has to be: Why must it be either? We, like chimps, are descended from a common ancestor, bonobos represent a further split from chimps on a separate branch of the hominoid lineage to us, some 2 million years ago. They have had as long as we have, 6 million years, to diverge from whatever that common ancestor was. It makes no sense to see either of them as some evolutionary template for our behaviour. We haven't the faintest idea what the common ancestor really looked like, nor how he behaved, and we probably never will.

In November 2005, Helene Guldberg came out with a hostile review of de Waal's book in *Spiked Science* on-line. Frans de Waal, she said, tells us we should get in touch with our 'inner ape'. 'Speak for yourself!' When she met de Waal to discuss his work she recalls that he mocked her for being anthropocentric—demanding that humans are a special, unique case. 'You are talking about the unique capacities of humans', he jibed, 'but genetically we are 98.5% identical to chimps and bonobos, and mentally, socially, and emotionally we are probably also 98.5% identical to chimps and bonobos. We love to emphasize the little difference that exists and cling to it and make a big deal out of it, but the similarities vastly outnumber the differences.'

De Waal's argument is, I believe, one of the most egregious fallacies in studies of evolution. It is to claim that, simply because we are closely related genetically, we must be closely related behaviourally. It is one form of the argument from analogy. The comparative cognitive psychologist, Daniel Povinelli, traces it back to the days of Darwin himself:

> If chimpanzees and humans share 98.6% of their genetic material, then doesn't it follow that there ought to be an extraordinarily high degree of mental similarity as well? This idea has been paraded so frequently through the introductory paragraphs of both scholarly journal articles and the popular press alike, that it has come to constitute an anthem of sorts; a melody, which if not sung, raises doubts as to one's allegiance to the cause of defending the chimpanzee's dignity.

For Povinelli it is an inherent fallacy deeply embedded in the history of psychology:

> It is an assumption that the 19th century philosopher David Hume found to be unassailable; it is an assumption that was present over a century ago when the field of comparative psychology was being founded by Charles Darwin; and it is an assumption that was even explicitly stated when the field was formally codified by Darwin's champion, George John Romanes. Even today, the invisible tentacles of this assumption run deep and tangled through our efforts to understand

the minds of other species. The assumption, quite simply, is that when it comes to trying to compare the mental lives of humans and other species, analogous behaviours imply analogous minds. The argument by analogy.

It was Romanes who developed Darwinian thought into the first comparative psychology in two treatises, *Animal Intelligence* in 1882 and *Mental Evolution In Animals* in 1883. Thus was psychology born with the argument from analogy, says Povinelli, and Romanes gives himself away by the following:

> Starting from what we know of the operations of my own individual mind, and the activities which in my own organism they prompt, we proceed by analogy to infer from the observable activities of other organisms what are the mental operations that underlie them.

The actor's grim motto is 'Never work with children and animals!' Comparative cognitive psychologists make a living doing both. When you design an experiment to reveal the thought processes of a child of three years or older you can be fairly sure of their reasoning because you can ask them and they can tell you. With tiny tots and all animal species you have to infer what they are thinking by designing fiendishly clever experiments and assessing their performance on them. The philosopher Thomas Nagel pointed out the problem when he asked 'What is it like to be a bat?' He answered, of course, that we may never know, perhaps we never can know. What is it like to be a chimp? Primatologists and cognitive psychologists have been debating this for over fifty years and are still no closer to an answer.

How do humans make sense of the world around them? How do chimps? Both species can learn from past experience that action A inevitably leads to outcome B—as in a dominant chimpanzee male witnessing a furtive copulation between a female and a sub-dominant male. Bitter experience soon teaches the sub-dominant that being spotted leads to a thrashing! Both species can make inferences as to how one thing causes another from the evidence of their own eyes, but decades of research into infant and childhood cognition has led to a deep understanding that we humans go far beyond the use of visual cues as to how one thing causes another to happen because we are able to make sense of the world by inferring causation by things we cannot see—things that are unobservable, invisible. Thus, when we observe another person doing something, we don't content ourselves with some mechanical description of what is going on—as if observing a machine or a robot—we understand that the other person *intends* to do something because he or she is being guided by internal mental states like desire, belief, and knowledge. We cannot see a desire and so we interpret the behaviour of others around us by inferring such mental states. In the words of one of the leading researchers in this area, Simon Baron-Cohen, we are 'mind-readers'. This does not mean that we are all Mystic Megs. It simply means that we naturally and effortlessly impute intentions to other people by inferring the mental states that we believe are guiding them. We are, if you like, forming a theory about why so-and-so is doing such-and-such by invoking the invisible. This is why this cognitive ability has become known as 'theory of mind'.

In the same way that we interpret behaviour in the social world by invoking mental states in others' heads, we have acquired the ability to predict the movement and interaction of objects by reasoning about physical forces, like gravity, which we also cannot see. Thus, we predict that when we drop an object it falls straight down to the ground, as, eventually, does something we throw up into the air; we predict that when we throw a stone with greater force than before, it will travel further before it hits the ground—or a prey animal; and that for one object to cause another to move it must either be in contact with it, or with intermediary objects that are.

These ideas of human cognition are sometimes hard to 'get' because we all take them so much for granted. We find it impossible to imagine a world in which we couldn't explain another's actions because we had no concept of 'she wants, she believes, she thinks that'. You try it. Understanding your partner opening the fridge door, only to close it again with a mystified or frustrated expression on her face would be impossible if you were unable to theorize: 'She has gone to the fridge because she *wants* some fresh milk and she *thinks* there is a bottle in there, but what she *does not know* is that we have used it all up and forgotten to buy some more!' Imagine how frightening and mysterious the world would be if another human approached you baring his teeth and you could not understand that he was grinning with joy at seeing you, not wishing to bite you! Many autistic children find facial emotion difficult to interpret, finding difficulty with joy, fear, anger, surprise, or sadness, never mind more complex emotions like embarrassment, pride, and shame. Simon Baron-Cohen calls them 'mind-blind'. What would you see, when you looked at another person, if you could not read emotional expression? The psychologist Alison Gopnik thinks you might just see a bag of skin, stuffed into clothes, draped into furniture, with two little black things up on the face! Is this what people with autism see? What do chimpanzees see? In the same vein, imagine playing a game of pool without the faintest idea that the cue ball must hit the red at a certain fine angle or you will not be able to pot it. Or imagine a world in which your understanding of time and space was so hazy that causation, one event causing another to happen, was impossible to grasp.

This form of cognition, by which we interpret both the social world and the physical world, is called 'folk psychology' and 'folk physics' because both are totally naive—they are totally uninformed by either physics or psychology in the text-book scientific sense. They are commonsense theories and may very well be at odds with proper scientific theories. For instance, some of us prefer to believe that when water exits a coiled hose-pipe it continues to describe a circle. Folk physics and folk psychology give us insights into physical and social causality—cause and effect—before, and even when, we understand something in a formal scientific sense about invisible forces like gravity and electro-magnetism and belief. Before we are able to describe the flight of a stone in terms of the transference of kinetic energy from arm to stone, and the friction of air molecules. The key point, in each case, is that we form theories which predict the behaviour of people or objects by reasoning about things we cannot see—invisible, unobservable things. Although this theory-building may be scientifically naive, according to the prominent American psychologists Andrew Meltzoff and Alison Gopnik, we go about it as if we were scientists—and we start doing it from day one. Gopnik and Meltzoff argue that little children, even tiny babies, act like little scientists, forming and discarding theories about the world around them. (Or, to put it another way, scientists are just big babies!)

Newborn babies, Meltzoff and Gopnik believe, have an innate ability to infer things about the world by changing and re-working an initial 'theory' about how the world works on the basis of experience. Meltzoff has shown, for instance, that babies as young as 42 minutes can imitate facial expressions: one baby successfully imitated him pulling his face and sticking his tongue out, for instance. Within the first year they learn to prefer human faces to other objects and the sound of the human voice to other sounds, and they learn to recognize a range of facial expressions and tell the difference between them and internal states like happiness, anger, sadness, and fear. They are mapping the outside to the inside. We all know the old saw about the eyes being 'the gateway to the soul', but psychologists believe they are the gateway to the mind. So do babies. One-year-olds start looking where other people are pointing—they know that this action is directed at something interesting in the outside world. They soon learn that they can engage the attention of another to a shared object of interest, by pointing and gesturing at it. They learn to empathize in that they begin to feel about something the way they perceive others are feeling about it. At this point, according to Gopnik, they are well aware of the concept of 'mind'. But, at this point, mind is everywhere, like the ether, and there is no dissociation between 'my mind' and 'your mind' or an understanding that the two might be different.

A few years ago I filmed Alison Gopnik running an experiment with a 14-month-old boy which demonstrated this beautifully. She sat at a table opposite the little boy and showed him two bowls of food—one containing goldfish crackers, which all kids love, and the other containing raw broccoli, which, of course, they intensely dislike. The little boy demonstrated his preference by cheerfully shovelling some crackers into his mouth. Then Gopnik sampled from each of the bowls in turn to show the boy what *she* did and didn't like. So, on picking up a cracker, she said 'Yuk!' and pulled a disgusted face, whereas on picking up a piece of broccoli she said 'Mmm! Yum!' and put on a gleeful face. Then she pushed both bowls to within reach of the baby boy, held out her hand right in the middle, between the bowls, and asked 'Can you give me some?' The little boy gave her some crackers even though she had shown that it was the broccoli that floated her boat—because he assumed that, because he loved crackers, everyone else must. Four months later he would not make this mistake, and Gopnik would get the broccoli.

So, by 18 months, a child has a basic level of understanding of hidden things like desires. Gopnik believes that the tantrums of the terrible twos are caused by the penny suddenly dropping that what the child wants is at odds with the desires of its parents. By the time a child is about four, she will exercise theory of mind at a more advanced level by understanding that other people have beliefs, and that, therefore, they may have false beliefs. There are several psychological tests to discover the developmental age at which this higher-order understanding of theory of mind kicks in. One test, used by Gopnik and her colleagues, involves a box of sweets. The child assumes that it will contain sweets but when she opens it, pencils fall out. Now, if you ask a three-year-old child what her friend will think is in the box, if she were to now enter the room and be presented with it for the first time, she will answer 'pencils' because she assumes the friend is able to share her new state of knowledge. However, by about four years, the child will answer 'sweets!' because she understands that the false belief will still be harboured in her friend's head, even though she knows differently.

How do children reason about cause and effect? From 3 years old they are capable of keen inference. This causal attribution comes early and is very strong. Alison Gopnik, and her then graduate student, David Sobel, invented an ingenious game called the Blicket Detector and demonstrated it in children as young as two. Children were shown four blocks, two of which were identified as 'blickets'. The experimenter then picked up one of the blickets and put it on top of a bogus machine—which was identified as the 'blicket detector'. A light went on at the front of the machine and it started playing a tune. This was, of course, operated out of sight by one of the experimenters throwing a hidden switch. Then the experimenter picked up a block that had not been called a blicket, put it on top of the machine—and nothing happened. When asked to identify another block that would start the machine even 2-year-olds invariably chose the other 'blicket'. They had used their causal powers to extend the same attribute to it because it shared the same name. Though they were never told that *only* blickets could make the light go on, they had inferred it. Some children even tried to take the 'blickets' apart to find out how they worked!

The $64,000 question—the holy grail for the student of human cognitive evolution—is whether or not these abilities are unique to humans, or whether we share these abilities with our nearest primate relative, the chimpanzee. After all, according to the argument from analogy, if we humans have powers of folk physics and psychology and chimps are 98.6% genetically identical to us—so, in large part, must they. Popular accounts of chimpanzee behaviour love to erode the idea of human uniqueness by pointing to recent discoveries in tool use, communication, and deception. But do chimpanzees have a well-developed 'folk' psychology, or 'folk' physics, when they make tools or interact with each other, or are they restricted to forming theories about causation based only upon things they can observe? If the latter, then human cognition is truly unique in the animal kingdom, and, crucially for us, it will dramatically distinguish humans from all the other primates. This matter has become vital to resolve.

It is far beyond the scope of this book to attempt a detailed overview of the history of chimpanzee–human comparative psychology. The literature is vast and research has accumulated over decades. In order to give an inkling of the problems and disputes involved I am going to take a narrow window of the sociology of science by concentrating on the work of two groups of scientists. One is Daniel Povinelli, who runs the Cognitive Evolution Group at the University of Louisiana, and the other is the Department of Developmental and Comparative Psychology at the Institute of Evolutionary Anthropology in Leipzig, headed by Michael Tomasello, Josep Call, and, until very recently, Brian Hare. Both groups research folk physics and folk psychology. Both groups are widely considered to be pre-eminent in their field. Both groups have veered from pessimism to optimism, or vice versa, concerning whether or not chimpanzees have theory of mind, or, more importantly, whether any of their experiments can prove it. And relationships between the two groups have deteriorated in the last few years due to Povinelli's robust scepticism in the face of experimental results the Leipzig group consider to be very reasonable proof that chimpanzees understand something about the concept of mental states. Povinelli's current stance is obdurate, clear, and unequivocal. No test to date, he asserts, has reliably demonstrated that chimpanzees (or any other primate for that matter) have any comprehension of the mental life of other individuals. The tests themselves cannot tell you if the primates concerned are acting purely on what they observe of the outward behaviour of others or whether they understand that they are engaging with the hidden contents of other minds. Likewise, tests to determine what chimpanzees understand of the invisible world of physical forces are equally equivocal. A new paradigm must be established, he says—new and better thought-out experiments—before anyone can consider that theory of mind has been proved in chimpanzees and the other great apes. Otherwise researchers are inevitably reduced to asserting or assuming that the behaviour they have recorded has been the result of mentalistic processes. 'He did this therefore he *must* have thought that ...' This is the gauntlet that Povinelli has thrown down, in no uncertain fashion, to the research community of which he is a part.

Povinelli's earlier research concentrated on trying to prove that chimps, like us, understand that *seeing* leads to the internal state of *knowing*; that chimps can take the perspective of another, giving them the ability to reverse roles in joint tasks; that chimps can discern between malice and accident such that they favour a trainer who appears to accidentally spill food over a trainer who deliberately spills it; and that chimps can accept pointing cues to the location of food. His results reflected his early optimism. But, by the mid-nineties, following a devastating critique of the whole field of comparative cognitive psychology by Cecilia Heyes of University College, London, he claims 'the scales fell from my eyes', compares himself to a drunken man sobering up, and his tone becomes more pessimistic. His landmark book, *Folk Physics for Apes*, published in 2000, remains by far the most comprehensive and dazzlingly inventive account of research into the extent to which chimps appreciate causation in the physical world around them. However, it attracted the ire of many in the comparative psychology community because it was published en bloc as a book and therefore never submitted piecemeal to a peer-reviewed scientific journal (though many other research papers germane to these issues have been), causing one prominent scientist in the area, Marc Hauser of Harvard University, to grumpily exclaim that, had he been on any review panel for much of this work, he would not have recommended publication! Povinelli went for the *coup d'état* and it irks!

The behaviour that best illustrates the principles of folk physics is tool use. Chimpanzees across Africa have been observed using twigs to dip for honey and probe for bees, termites, and algae; have been seen to use leaves to sponge up liquid; crack nuts against hard surfaces, or by using crude stone hammer and anvils; pound nuts as if using a crude mortar and pestle; and use simple hooks, found in vegetation, to forage for insects. There is quite a lot of cultural variation in these behaviours—not all colonies of chimps from different parts of Africa exhibit all behaviours—and the big questions are: How much do chimps understand about why a hooked twig will retrieve insects from the rotten bark of a tree? Do they know why a stone will crack a nut when dashed down onto it or why a T-shaped stick will pull food toward them, like a rake? Have they selected and shaped their tools because they are aware of the physical forces involved and understand something about causality, or by blind trial-and-error? Mere observation of chimpanzee tool use in the wild cannot resolve these underlying issues, which is why Povinelli and many other cognitive scientists have taken tool use into the laboratory to try to peep into the mind of the chimpanzee.

One of Povinelli's first experiments was inspired by his then student Jim Reaux. Both were, of course, well acquainted with Jane Goodall's graphic descriptions of chimpanzees using twigs, pieces of bark, and grass stalks to probe into the narrow holes in termite mounds. The technique is effective because the food emerges clinging to the tool with its vice-like mandibles. But do chimps have some notion of causality which guides them to the appropriate selection of width and length in the tools? To find out, they used the 'trap-tube' methods first introduced by Italian researcher Elizabetta Visalberghi for capuchin monkeys and chimps. Here a peanut is placed in a clear perspex tube and the animal has to dislodge it by poking at it, from one end or another, with thin sticks of various lengths, a bundle of sticks bound together, or a stick from which cross-joints have to be removed. The monkeys eventually learned the task, by selecting long sticks, or by shunting three short sticks together, or by unpicking the bundle of sticks before use, but they never did conquer the task where they had to remove cross-joints (analogous to pruning side branches off a twig in the wild). Her chimpanzees fared little better. Although they unbundled the sticks without any errors they, like the monkeys, appeared not to understand the stick that had two cross-pieces (like a telegraph pole), one at each end, often taking one cross-piece out but then trying to insert the wrong end into the tube.

In order to further probe whether or not the primates had any genuine insight into their choice of stick, Visalberghi had created an ingenious elaboration of the experiment whereby the tube now had a downward pointing side-arm which operated as a trap. The peanut was placed at random either slightly to the right or left of it. Would the monkeys learn to push from the appropriate side to avoid the peanut falling into the trap? Only one monkey was successful but the question still unresolved was whether it had solved the task by operating a simple and inflexible rule like 'poke for the reward through the end of the tube furthest away from the food' rather than realizing anything about trap avoidance. To test this, they simply inverted the tube. Now the trap was still there, but pointing upwards and therefore totally ineffective, while the peanut was still placed either to the left or to the right of it. The monkey could dislodge the peanut by pushing from either side, but it stuck rigidly to pushing from the opening farthest away from the reward.

Povinelli set out to repeat the trap-tube test with his troupe of seven chimps. Of the four who survived the opening rounds, only one—Megan—performed above chance in avoiding the trap by pushing from the appropriate end of the tube. However, even Megan performed only at chance for the first 50 trials! When they inverted the trap Megan continued to follow the procedural rule, 'push from the side farthest away from the food' 39 out of 40 times, even though the trap was now irrelevant. In a third test they placed the rod on the side of the inverted tube nearest to where the food was.

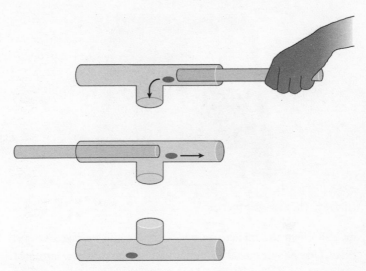

FIGURE 2. The trap-tube problem

Megan insisted on the procedural rule by picking up the stick, if necessary, and taking it round to the end of the tube furthest away from
the food before pushing. All this suggested very strongly to Povinelli
and Reaux that she was following a low-level rule, even though it
involved transport costs in one irrelevant condition, and that she had
no inkling of the causal role of traps. They dreamed up several other
combinations of trap, bait, and stick to probe Megan's insight further,
the clincher being a condition where the tool was always pre-placed
in the left-hand side of the tube with the food either near- or far-side
of the inverted trap; near-side of the functional trap; and far-side of
the functional trap. The version where the trap was functional, the
tool was pre-placed on the left side, and the food was also on the left
or near-side of the trap, was the critical one. If she had understood
about traps she should have removed the stick and re-inserted it in
the opposite side. She never did.

To probe this kind of tool use a little further they conceived a
'trap-table' problem. In this task the apes were presented with a table,

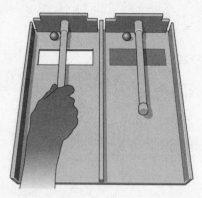

FIGURE 3. The trap-table problem

divided down the middle, on which food could be raked toward them. On one side the raked food would have to pass over a painted rectangle, on the other it would be trapped by a rectangular hole. Only Megan—from trial 1—scored highly by selecting to rake across the painted rectangle, as opposed to the surface cut into by a hole, first time, on 80% of the trials. The others did not score above chance.

Nevertheless, chimps in captivity have frequently been observed to use sticks or rakes to pull out-of-reach food toward them—which suggests chimpanzees do understand something about how rakes work. However, Povinelli's chimps couldn't distinguish between the efficacy of a rake held in the normal position—a bit like a snooker bridge—and one in an inverted position, to capture a cookie on a table. They consistently failed to understand that to move the cookie they had to make contact between the two by using the inverted rake. Povinelli concluded that the apes learn to appreciate the perceived visual proximity of tool and reward but not the concept of physical contact necessary to transfer force.

In the wild, chimps are successful at dipping for termites by inserting appropriate-diameter grass-stalks and twigs into holes and carefully removing them without brushing the termites off. This suggests

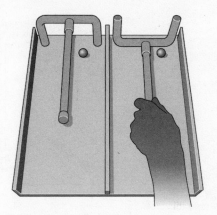

FIGURE 4. The inverted rake problem

they may have an explicit understanding of the appropriate size and configuration of the end of a tool. Povinelli's team tested this using an appropriate tool, which was straight at one end but had a side-bar at the other, versus impossible tools which were shaped like a T at each end or were rectangular bars, which the chimps had to choose between to poke free an apple through a hole in a plexiglass sheet. They showed only mild preference for the appropriate tool over the impossible one, though they often tried to employ the difficult end of it. The chimps' performance did not justify a high-level interpretation that they were understanding the causal structure of the task. They seemed to be more influenced by the grasping qualities of the tool—what they could use to handle it with—than whether it had the ideal straight end for insertion into the hole and poking the apple. The straight end provided the better handle for them at the expense of easy poking with the other end.

Povinelli ploughed on to further research the question of whether or not apes understand that for one thing to cause another to move there must be physical connection. He began with a series of strings and object tasks in which several strings were shown in different geometrical relations to the reward—a banana—but only one was

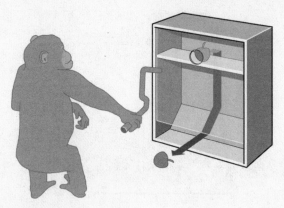

FIGURE 5. The tool insertion problem

attached to it. The banana would move only in two conditions, one where it was tied, and one where it lay on top of the rope (provided that this was gently pulled). If either of these conditions were appreciated, an understanding of physical connection would be implied.

However, the only condition where the apes demonstrated a significant preference for one orientation over another was when they had to choose between a rope tied to a banana versus one that ended short of it. Did they understand anything about the difference between a string physically tied to the banana or one that merely appeared to be in contact? In all other tests it seemed as if they were more impressed with the perceptual degree of contact rather than actual physical contact—the optics of the situation seemed more important to them than the physicality of it. Had they understood deeply about connection they should have chosen to pull a string that lay under a banana versus a string draped over it. But they didn't.

Chimps have been seen to modify tools in the wild. For instance, lateral twigs are pruned off small branches so that they can be pushed into holes; bark is sometimes removed to decrease the diameter of twigs; and ends of twigs or grass stalks might be chewed so that they splay out giving greater purchase to the termite quarry. But what do

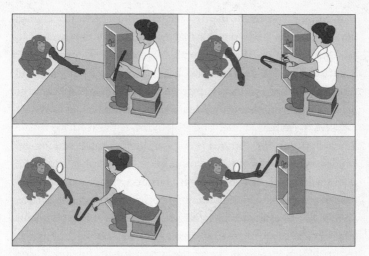

FIGURE 6. The tool modification problem

chimps explicitly know about these adaptations? Are they just running scripts learned through imitation, or feedback from a myriad of their own attempts?

Povinelli's researchers tested this by giving chimps a tool of bendable piping with which they were thoroughly familiarized. In the experiment, a researcher would either bend the pipe into an S or a C and then give it to the chimp, whose task was to bend it into something approaching a straight pipe that would be effective for the task. Only two chimps made any attempt to modify the tool and they did it in only one trial each. The researchers tried to teach the chimps by demonstrating how to straighten out the bent tubes; however, they couldn't catch on. Povinelli concluded that chimps have no explicit causal understanding of how the shapes they create physically interact with the world. It is all trial-and-error learning. This may be why chimpanzee material culture, although sufficient for their way of life, and impressive from the point of view of other non-human primates, is so very limited compared to that of our own, he says.

According to Povinelli, there is precious little evidence that chimps have a grasp of 'folk' physics that is in any way comparable to our own. How do they fare with respect to 'folk' psychology? Primatologists for decades have been recording fascinating and thought-provoking incidents in the wild where primates appear to be deliberately manipulating the behaviour of other individuals. We are indebted to Andrew Whiten and Dick Byrne, of the Scottish Primate Research Group, for winnowing out many of these observations and condensing then down into a major book, *Machiavellian Intelligence*, published in 1988 and followed in 1990 by a comprehensive data-base of reports of tactical deception in primates. Social intelligence, as opposed to estimates of intelligence gleaned from experiments where primates are required to manipulate tools and apparatus, has received far more attention over the last forty years because Alison Jolly, and then Nick Humphrey, argued that, in the wild, primates face very few gadget- or technology-based challenges and that the sternest challenge they face comes from other members of the same species. Thus, selection for how to handle other primates—social skills in cooperation and competition—were much more likely to have been the driving force for the evolution of primate intelligence than physics.

Whiten and Byrne interpreted deception as a social tool to gain advantage without confrontation, and the deceived as social tools to the deceiver's ends. By 1990 they had amassed 253 records which they divided into different categories of deception and ranked by the extent to which the reporter felt that the behaviour he or she had witnessed could best be interpreted by appealing to less complex forms of learning like trial-and-error and reinforcement, or more complex cognition like 'mind-reading'. In other words, was there evidence that the primate concerned 'knew' why the deception was working?

Many of the records involved concealment. For instance, a female hiding behind a rock in order to enjoy sneaky grooming or copula-tion with a subordinate male, safe in the knowledge that she was out of sight of the dominant male, or a male chimpanzee demonstrating his erection to a female while concealing it from the dominant male. Others involved distracting the attention of other individuals either by 'pretending' to fix the gaze on some spurious distant object of interest, or by distracting the attention of another from where food was actually concealed by wandering off and attending to spurious locations, or by feigning lack of interest. Thus, many examples involved understanding something about the reasons for the orientation of other individuals—and have given rise to perennial arguments about what the primates actually know about looking/seeing/knowing that have endured to this day. For instance, the knowledge that if a dominant is oriented toward you, retribution is likely to follow, versus understanding that the dominant can see you *and therefore knows what you are up to*, therefore retribution is likely to follow. Again, as for tool use, no observation in the wild can reliably distinguish between the lower-order and higher-order interpretations of deceptive behaviour, and although Byrne and Whiten piled anecdote high upon anecdote, sheer volume of records got them no nearer to the truth, though some were more persuasive than others.

Any account claiming intentional deception on the part of a primate depends on accepting that not only could one individual conceive that another was *looking* at something when orienting in a certain direction, but that he could therefore *see* that certain something which led to the invisible mental state of *knowing* something about it or forming a *belief* about it. Povinelli set out to create explicit laboratory-based tests of Byrne's and Whiten's field evidence for mind-reading. Early testing, in Povinelli's laboratory, and those of others, had shown that chimps behaved just like 18-month-old children in following the gaze of an experimenter. They responded when both the head and eyes turned to look at an object, or when the eyes alone moved. But were they aware this meant the experimenter could see? A chimpanzee might follow the gaze of another so that they both end up looking at the same thing without an understanding of attentional states coming into it at all. Or it might turn and look where you are looking because it wanted to see what you see—a higher-order level of cognition.

Povinelli tested whether or not the chimps could distinguish between someone who could see them and someone who could not. Could they understand that their begging gestures needed to be seen in order to be effective? Would they prefer to beg from someone whose eyes they could see and who was looking at them? After all, their favourite game was to up-end buckets, bowls, sacks, or card-board boxes over their heads and then walk around the enclosure, bumping into things. Hard not to assume that they understood *they* could not see. But when one of two experimenters either wore a bucket, a blind-fold, or held their hands over their eyes, the chimps showed no preference in whom they begged to. However they *did* discriminate between an experimenter facing front and one with his back turned. Why the discrepancy? Perhaps the back–front distinc-tion was simply the easiest case of seeing and not seeing and one the chimps mastered? Or was it that they had simply learned, in the pre-trials, to enter the chamber and gesture to an experimenter facing them? Or perhaps they were simply following a rule to ges-ture to the front of others without any appreciation of 'seeing' at all? This was tested by a 'peeking' or 'looking over the shoulder' test, where one of the experimenters faced away and the other looked over her shoulder. The chimps showed no preference. They seemed to need the full-frontal mode. Neither did they discriminate between an experimenter holding a screen over her face versus one who did not, nor an experimenter distractedly looking away to one side versus one looking to the front. In fact, in this test, they followed the distracted gaze to the corner of the room but still did not appear to twig that begging to her would be pointless. When a blindfold was put over the eyes of one experimenter and the mouth of the other, again there was no discrimination. Neither did they spontaneously discriminate between an experimenter whose eyes were shut versus one whose eyes were open.

How then to explain the clear pleasure that chimps get from play behaviour when they deliberately occlude their vision by putting things on their heads? If they do not know that they cannot see, what is going on? Perhaps, Povinelli reasons, the experience of visual occlusion is not represented any more explicitly than the sensation of scratching relieving an itch. It could also be that they understand their mental states but cannot extrapolate that understanding to the idea that other individuals may have them. They ran the test on 3-year-old children, who do not know that seeing is knowing until they are 4. The children performed almost flawlessly, suggesting an organism only had to compute that seeing meant attending in order to pass these tests. Had the chimps fallen at a low hurdle after all?

The problem was resolved some time later, when the chimps were re-tested on 'seeing/not seeing—the eyes open/eyes closed' protocol and the 'screens covering face' condition. Despite several years previously having posted high correct scores after learning through the trial, they scored no higher than chance. And this having been tested, in the intervening years, on all sorts of aspects of attention. Specific rules had been forgotten and the apes had clearly failed, in any way, to generalize from other tests on similar cognitive abilities. The various tests on aspects of attention had failed to coalesce, in the chimps' minds, into a central construct.

However, these tests had now left the experimenters with a disparity in that, in earlier gaze-following tests using barriers separating human experimenter and chimpanzee subject, the chimps behaved as if corresponding to a high-level cognitive explanation in that they looked round the barrier to 'see' what the experimenter was looking at, rather than extrapolating a line through the barrier to somewhere behind them. Yet, in the 'seeing/not seeing' tests they had demonstrated a low-level cognitive ability. To try to resolve this, Povinelli used a test where two opaque cups were placed either side of a table top. Food was hidden underneath one of them. The experimenter either gazed *at* the correct cup or *above* it. Could the chimps discriminate the correct cue? Three-year-old children chose correctly when the gaze was fixed on the cup but at random when the attention was above the cup. The chimps were much less discriminating, following the gaze to select a cup whether the experimenter was looking at it or above it—the general direction seemed enough. Despite their excellent gaze-following abilities, concluded Povinelli, they still did not understand that gaze is related to subjective states of attention. So, what is it with the barriers? It may be, says Povinelli, that the chimps do not understand that there must be something behind the barrier that is occupying the experimenter's attention, but that they have generally merely learned that in this situation the space on their side of the barrier is irrelevant.

In 2003, the comparative cognitive psychology group at the Max Planck Institute for Evolutionary Anthropology in Leipzig—Michael Tomasello, Josep Call, and Brian Hare—reflected back over several years of cognitive tests with chimpanzees. As recently as 1997, they explained, in a comprehensive review of all the evidence from a number of laboratories, they had concluded that non-human primates understand much about the behaviour of other members of their own species but nothing about their psychological states. They agreed 100% with Povinelli that non-human primates cannot deal cognitively with unobservable causes at all.

However, they said, new data had emerged in the intervening five years that 'required modification' of the rather black-and-white hypothesis that chimps understand no psychological states whereas humans over the age of four do. Telling chimps and humans apart, they said, had just got harder. They had seen light in Povinelli's gloom thanks to what they called a series of 'breakthrough experiments', mainly done by Brian Hare, and reported in 2000. Hare had worried that the main reason for the negative results pouring out of his and Povinelli's labs might lie not so much in the chimps' inability

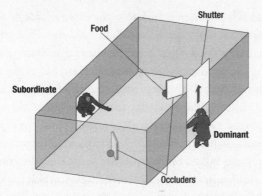

FIGURE 7. Brian Hare's food competition experiment using barriers

to reason about mental states as in the very artificial nature of the experiments used to test for it, which, he argued, ran against the grain of chimpanzee behaviour in the wild, where they compete for food resources and are unused to an individual being cooperative and offering them something. Perhaps, he wondered, all the previous experiments lacked street credibility for the chimps. They simply didn't understand the social context.

He decided to create an experimental set-up where he could frame the 'seeing-knowing' question in the context of competition for food between a dominant and a subordinate chimp. The two chimps were separated by screens, which could be raised or lowered, from a central chamber which contained two barriers made out of opaque cloth bags. Both animals were allowed to watch while the experimenter placed two pieces of food in the central chamber, sometimes in the open and sometimes behind one of the barriers. Then both chimps were allowed into the chamber. The dominant, of course, took all the pieces of food it could see, and if both pieces had been placed in the open, it grabbed the lot. Occasionally, however, the experimenter had placed a piece of food behind one of the barriers in a way that was difficult for the dominant chimp to monitor. The question was whether or not the subordinate would preferentially go for the food only it could see, because it had been placed on its side of a barrier. The subordinate was given a very slight head start to encourage it to compete with the dominant and to ensure that it was not simply reacting to the dominant's behaviour on release, but on the remembered 'who saw what?' condition when the chamber was baited.

Their measure of the choice the subordinate made was based on the completeness of its approach to the correct barrier: no approach, half approach (half-way toward the barrier), and full approach, when the subordinate crossed the half-way line in the right direction. Note (because this becomes a major Povinelli criticism) that the measure has nothing to do with whether or not the subordinate arrives at the food, reaches out, and grasps it.

Their results showed that subordinates did indeed go toward the food that only they could see much more often than the food that both they and the dominant could see (because it was either out in the open or the dominant had seen it placed behind a barrier). This suggested to Hare that they did understand something about seeing and not seeing. But did they understand anything about seeing being equivalent to knowing?

In a follow-up experiment the subordinate was always allowed to witness one piece of food being placed on its side of one or another of the barriers. In the first condition, the dominant was also allowed to witness the action; in the second condition the dominant was not allowed to see and was thus ignorant of food location; and in a third condition the dominant was misinformed about the food location in that he was first allowed to watch the food being placed, but the food was switched while his screen was subsequently lowered. The screen was always completely lowered on the dominant's side for a few seconds before the subordinate, then the dominant, were released into the chamber. If the subordinate had remembered whether or not the dominant was looking, in the past, at the crucial moment when food was either placed or switched, and if he could thus compute that looking was akin to seeing (was akin to knowing), he should approach and retrieve the food more often when the dominant was either uninformed or misinformed about where it was, and thus avoid a bashing. He did.

In a variation of the experiment, the dominant who had witnessed where the food was placed was occasionally replaced with another 'surprise' dominant who had not witnessed anything. Could the subordinates further comprehend that the second dominant must be completely ignorant and therefore approach the food whenever she appeared? As predicted, the subordinate retrieved a larger percentage of food in the switched-dominant condition. They concluded that their results 'represent the clearest demonstration to date that chimpanzees know what individual group-mates do and do not "know", that is, what individual group-mates have and have not seen in the immediate past'.

Hare, Call, and Tomasello stopped short of concluding that the experiments had proved chimpanzees have a full-blown 'theory of mind' but they believed they had demonstrated that chimps have at least a limited understanding of perspective-taking because they use a 'rewind strategy' whereby they note what their competitor sees or does not see at the time of food placement and 'rewind the tape' in their heads at the point of decision as to which food to grab on release. In film terms this means that they cut to the other individual's POV. In this view, they concluded, chimpanzee social cognition is based on a representational understanding of the behaviour of others, which permits them to do things like remember, foresee, and communicatively manipulate the behaviour and social relationships of others.

The head of the department, Michael Tomasello, was quite clear at the time that there was still no evidence from anywhere that chimps understood anything about what others believe. They would, he said, be incapable of intentional deception, because that would require them to deliberately sow false beliefs in another's mind. We will never get anywhere in fine-tuning our knowledge of what components of theory of mind are available to chimps, and which are not, he argued, if we continue to treat theory of mind as some monolithic cognitive entity. Young children, for instance, below the age of four, seem to understand some things like seeing and knowing, and perspective, but they do not pass the classic tests of theory of mind—showing they understand that another person might harbour a false belief about something—until they are older. Perhaps chimps are cognitively arrested at the same level of social cognition as very young children?

> The stakes here are large. At issue is no less than the nature of human cognitive uniqueness. We now believe that our own and others' previous hypotheses to the effect that chimpanzees do not understand any psychological states at all were simply too sweeping. We believe that the way forward in research on chimpanzee social cognition is to 'turn up the microscope' so as to see which of the many different kinds of primate psychological states chimpanzees are able and not able to comprehend, and in what precise ways.

Rosalyn Karin D'Arcy and Danny Povinelli immediately set out to try to replicate Hare's experiments with the opaque bag barriers. They began by demolishing Hare's criteria. They found the retrieval of food measure to be highly misleading. In Hare's account, he fails to make clear, they say, that on the occasion when a piece of food is placed behind one of the barriers another piece is left in the open. This, they argue, is a crucial omission because if, on the conditions when the dominant has witnessed the placing of food, the subordinate allows him to pick up the food left out in the open, or avoids it, then the subordinate has no choice as to which food to take—only the food behind the barrier is left. Hobson's choice. They also find the measure of approach highly ambiguous and vague and substitute it with their own measure which requires the subordinate not only to approach the food but to reach out and touch it. Their experiments, very similar in concept and execution to those of Hare's, provided no support for Brian Hare's conclusions.

However, Povinelli had a much more fundamental objection to the Leipzig group's conclusions—that some element of theory of mind had been demonstrated. Even allowing the Leipzig researchers their results, he argued, there was still no way that the experiment could distinguish between an animal that understands that looking means seeing means knowing, and one that simply notes that the dominant was oriented toward the food at the time the human experimenter was placing it in the chamber. It is the interpretation of the researchers that puts the theory-of-mind gloss on it. 'The subordinate must have a concept of seeing and therefore knowing', they seem to be arguing. 'Are you convinced?' Povinelli is not convinced. For each occasion when a chimp *might* have been reasoning about unobservable mental states of others, in this case seeing and knowing, he might merely have been reasoning about an observable behaviour—in this case orienting toward—and that would also have been sufficient to allow him to behave appropriately. Povinelli makes his point graphically by using a thought experiment to suggest what might have been going through the subordinate's mind. In the following scenario the observable behavioural abstractions are in normal typeface and further reasoning that could only come from a theory of mind is italicized:

> 'He was present and facing the food when it was placed where it is now <*so he saw the food placed and currently knows where it is*> therefore he is likely to go after it'. The italicized section adds no further unique understanding of a mental process that could explain on its own whether or not the subordinate approaches the food barrier, since the first part of the statement would work perfectly well as a behavioural rule based on past experience and an explanation of the subordinate's actions. The same redundancy of mental state attribution is seen in the following: 'He was not present when the food was placed where it is now <*so he didn't see, therefore he doesn't know ...*>'

Can additional controls help? Hare used a similar set up in which one dominant, Joe, watched the baiting of food, but then when the door opened it revealed another dominant, Mary. The author's *own* theory of mind interpreted the experiment as follows: if the subordinate is more willing to approach the food when the new dominant is present, they must be reasoning, 'Well, Joe *saw* the food being placed so he *knows* where it is ... But look, it's Mary! She *didn't see* the food being placed, therefore she *doesn't know* ...' But, of course, that ignores that an intelligent chimpanzee could simply use the behavioural abstraction 'Joe was present and oriented; he will probably go after the food. Mary was not present, she probably won't'

The two research groups agreed to disagree. The Leipzig group preferred to believe that theory of mind was not a monolithic cognitive process; that chimpanzees, while not capable of full-blown theory of mind that would allow them to conceptualize 'beliefs', nevertheless possessed lower-level elements of theory of mind, such as being able to comprehend that if someone was looking at something, they had a subjective mental state of 'seeing' it, and probably a further mental realization of 'knowing' something about it. In that sense, chimpanzees could be viewed as cognitively arrested humans akin to the developmental stage of 2- to 3-year-old children. Povinelli and his colleagues preferred to believe that both chimpanzees and humans are capable of reasoning (and planning their behaviour) with respect to the outward, observable, behaviour of other individuals, but that humans might very well be unique in the animal kingdom in their peculiar 'add on' of being able to reason about unobservable mental states. They further do not like the 'arrested development' idea of chimpanzee social cognition.

'Maybe', Povinelli says, 'the human mind has evolved a unique mental system that cannot help distorting the chimpanzee's mind, obligatorily recreating it in its own image?' In other words because *we* have theory of mind we can't help but describe behaviour of non-humans (and sometimes inanimate objects) as if they too were aware of the beliefs, desires, and intentions of others. He's got a point. We humans are almost sick and delusional with anthropomorphism, we simply cannot help but ascribe mental states to our pets: 'My dog understands everything I say!'; our plants: 'That rambling rose has got a mind of its own!' Our literature is chock-full of 'little engines that could' and sapient trees who consort with imaginary dwarf hominids to bring about the downfall of wizards. My personal favourite is Tom Hanks' character in the feature *Cast Away*. In danger of losing his mind through loneliness, when cast up on a desert island, he retrieves a volleyball from the FedEx packages washed ashore from the plane, and, because the brand-name is 'Wilson', the ball, with a painted face, becomes 'Wilson'—his Man Friday!

So, if human psychologists have a dangerous propensity, because of the way our brains work, to attribute mental states to things whether they have them or not, we will always be biased toward seeing theory of mind as the most economical, even parsimonious conclusion to be drawn whenever experiments remotely merit it. Indeed it may seem the commonsense solution. Povinelli's answer to what he perceives as an anthropomorphism trap is to look to a much better experimental paradigm, and he does this by adapting an idea proffered by the British psychologist, Cecilia Heyes, back in 1998, that he quibbled with at the time.

Fit two buckets, he suggests, with visors, one of which is see-through and one of which is opaque. Let one be colour-coded by a ring of red paint, say, and the other by a ring of blue. This is the 'arbitrary' cue because, otherwise, from the outside, it is impossible to tell the difference. Then let the chimpanzees play with the buckets so that they have a chance to form subjective experiences of blindness when they wear the bucket with the blue-rimmed visor, and normal sight when they wear the bucket with the red-rimmed one. Then let them beg from two experimenters, one of whom wears the bucket with the red-rimmed visor, one the blue. If the chimps preferred to

beg toward the experimenter wearing red this would strongly suggest they had been able to transfer their first-person experience to a
second person and begged to the experimenter they reasoned could
see them.

Povinelli's colleague Jennifer Vonk has since tried out this
buckets-and-goggles experiment with chimpanzees brought up
in the enriched environments of human families—enculturated
chimpanzees—and they fail. Andrew Meltzoff, the University of
Washington child developmental psychologist who featured at the
beginning of this chapter, has tried an adaptation of the same experiment on 18-month-old children and they pass.

The Leipzig group, however, have flatly rejected this approach
because, in their opinion, it has 'very low ecological validity'. To
this day they have pressed ahead with an ambitious programme of
barrier-type experiments in which the chimps have to compete for
food with a human. Here the experimenter sits in a plexiglass booth
monitoring two dishes of food that can be reached by the chimps
via holes cut into the plexiglass sides. The experimenter's view of the
approaching chimps is restricted in a number of ways: by inserting
different combinations of occluding panels to left or right; by totally
occluding the booth and building two tunnels, one clear, one solid,
between the side holes and the food dishes, between which the chimp
must choose to thrust his arm; and even equipping one hole with
a rasping flap to produce a give-away noise should a chimp fumble
through it. Would the chimps choose the option that suggested they
were reasoning that the human inside the booth could not see—
projecting the situation to her point of view? They produced mod-
estly positive results and claim, though these experiments have yet to
be replicated by other groups, that the behaviour of the chimpanzees,
by taking advantage of the human's restricted sight to filch food,
amounted to *intentional deception*—a capacity they had ruled out for
chimpanzees only two years previously. However, they also accept
that the chimps may actually be withholding information from the

experimenter rather than deliberately sowing false beliefs—a more modest achievement.

Povinelli will have none of it. 'Same old, same old', he says, every higher-order explanation of chimp behaviour can be substituted to good effect by a lower-order explanation. So the two groups are at impasse. Over the years there has been increasing objection to the idea that humans are cognitively unique in the animal kingdom. Povinelli seems to be reversing this trend by asserting that humans decipher the social world around them by reading behaviour—just as chimps do—but have the evolutionary bolt-on, entirely idiosyncratic to our species, of theory of mind: the ability to read unobservable psychological states. The Leipzig group are more wedded to the idea of a biological continuum—befitting two species that are so close together genetically—where there is no black-and-white but shades of grey. In one view there is an unjumpable gap in cognition between the two species, and in the other, the cognition of chimps is rather like 'the darling buds of May' that eventually, over 6 million years of evolutionary time, flowered into full-blown human cognition.

To some extent the blame for this research impasse rests with the exclusive use of chimpanzees—and measures of chimpanzee cognition—to chart the mental evolution of our own species. So, to the extent that chimps may have at least some capacity to reason about mental states, we accept that as a clue to human cognitive evolution. To the extent to which they are eventually overtaken by the cognitive powers of developing children, we grant them the status of the 'nearly man'. It's been going on since Darwin's day.

Povinelli and the Leipzig group have locked antlers. Is there any way to free them? To make a clean break from the argument from analogy means to break the exclusive assumption that closely related animals have closely related brains giving rise to closely related cognition as evidenced by closely related behaviour. And the only way to do it is to widen the lens, put more and more species under the microscope, and break this monotonous exclusive comparison of chimpanzee and human. One begins to see the evolution of cognition in an entirely different light. In fact, despite their academic rumpus with Povinelli, the cognitive scientists at Leipzig have already cast their net wider with fascinating results. They began with experiments comparing chimpanzees with dogs.

Because of his very positive conclusions about chimpanzees' ability to follow gaze around barriers and to understand whether or not another's perspective is blocked by a barrier, Brian Hare was puzzled by chimps' relative failure on a test they should also have passed—called the object choice task. Here, food is hidden in one of several opaque containers and a human looks or points to the correct food location. Children, Hare reports, find this task trivial from 14 months onwards, but chimps find it beyond them. What about dogs?

> Give domestic dogs a crack at it and they show impressive flexibility in solving the same problem. The dogs were able to use several different behaviours to locate the hidden food at above chance levels: a human pointing to the target location—including distal pointing in which the experimenter stands over a metre away from the target and points in its direction using her cross-lateral hand; a human gazing to the target location, where the dog sees either the head turn or a static head looking toward the container; a human bowing or nodding to the target location; and a human placing a marker in front of a target location—a totally novel communicative cue. The dogs were even able to do the task correctly when the human walked toward the wrong container while pointing in the opposite direction at the correct one. In addition, dogs performed equally well whether cues were provided by other dogs or humans! They performed effectively from the very first trials—no learning was involved.

Dogs' preference for eyes and gaze is shown in the fact that they run around to place a retrieved ball in front of a human if he has turned his back to them; and they prefer to beg from a human whose head and eyes are visible and not covered by a blindfold or a bucket—something chimps cannot do spontaneously. They are also less likely to approach food when a human's eyes are open as opposed to being closed—again something chimps do not do. Moreover, reports Hare, dogs avoid approaching food when they are behind a large barrier but the food is in a window in the barrier with the human on the other side. This, even in a position where they cannot see the human and the human cannot see them.

Dogs' stunning ability in following social cues does not, however, apply across the cognitive board. Their brilliance is domain-specific. Dogs fail means-end tasks which require them to avoid pulling at a string which is not attached to food in favour of one that is connected—a task some primates are capable of. And they are dismal at following non-social cues as to the location of food—for example, discriminating between a board laid flat and one up-ended as if food is lying beneath. Their cognition is a peculiarly social specialization. They latch onto human eyes with laser-like accuracy.

Now, says Josep Call, senior researcher at Leipzig, do a four-species comparison taking wolves, chimps, dogs, and children. Chimps are not like humans, even little ones: they are neophobic, averse to touching new objects. Instead of learning from others they try to solve a problem on their own. Wolves are the same. In a test devised by Adam Miklosi, of Eotvos University, Hungary, wolves persisted, solitarily, with a task involving opening a sealed box, until the experimenter terminated the trial. Dogs are no better than wolves at opening boxes but they have the sense to give up early on and quite literally drop the problem in the human's lap! Now, children do the same thing. So here we can see, says Call, that closely related species are showing interesting differences. The wolf pairs with the chimp and the dog pairs with the child. Even though the dog and the wolf are closely related, some of the things that dogs do are closer to humans than to wolves, and even though the chimp and the child are closely related, chimps are more akin to wolves.

Now, no one is suggesting that dogs are more closely related, genetically, to humans than apes. And no researcher I know of claims that dogs are capable of full-blown theory of mind. Yet, here they are, matching, and occasionally out-performing, chimpanzees on a range of social cognitive tasks attending to, or getting the attention of, humans. Clearly their human-like social cognitive skills have nothing to do with genetic proximity to us. Adam Miklosi studies dogs and their owners in naturalistic settings. This heightened social cognition in dogs, he says, has been brought about by their entrance into human society—living cheek by jowl with humans—and where their very livelihood depended on the acute monitoring of human cues and actions. Dogs should thus adapt to life with humans by evolving novel social behaviours that are *functional analogues* of human behaviour—convergent traits. This convergence of behaviours doesn't necessarily imply the same cognitive mechanisms are at work. In the same way that PCs and Macintosh computers will allow you to do the same jobs, but with different software running on different silicon architecture, dogs' brains have converged on a suite of behaviours that, over a limited range, work in the same way as some of the component skills of human theory of mind—like gaze-following and an understanding of attention. They are a prime example of how to buck the argument by analogy.

By introducing more species to the picture the argument from analogy, or the argument from genetic proximity, looks a great deal less secure. In the next chapter I want to take this idea of convergent evolution to much greater extremes, which, I believe, completely blows the argument from analogy apart. The species involved diverged from us and the apes several hundred million years ago—they are birds.

CHAPTER 9

Clever Corvids

In the last chapter we tried to find a way out of an impasse generated by the very narrow, blinkered approach to comparative cognitive psychology that has placed undue emphasis on comparisons between chimpanzees and humans. Outside this chimp–human comparative world are a breed of comparative cognitive psychologists who despair of this myopic bias and have proved their point with compelling psychological research on birds, a group of animals far removed from humans and the rest of the primates because they diverged over 280 million years ago. Chief among these scientists are the husband-and-wife duo from the University of Cambridge, Dr Nathan Emery and Professor Nicola Clayton. How did comparative psychology become so primatocentric? So chimp-ist? How come primates have achieved such special status as lowly near-humans? Nathan Emery:

> We believe that this harks back to the beginnings of prima-
> tology as a discipline. Early studies of primates in zoos or the
> laboratory were performed to determine whether the cognitive
> abilities of primates could be compared to the cognitive abil-
> ities of humans (stemming from the evolutionary arguments

of Darwin and others). This approach has continued unabated since, culminating in work that was concerned with whether chimpanzees have the capacities for human language, relational learning, and Theory of Mind. The tradition of comparing primates with humans on tests of cognition continues to this day.

Field research also had a hand to play in promoting primatocentrism because most of the work was done by anthropologists who wanted to get a handle on how extinct humans might have behaved. This is the very reason why Louis Leakey sent Jane Goodall, Birute Galdikas, and Dian Fossey into their relative jungles to study the three great apes, chimpanzees, orang-utans, and gorillas, respectively. Later, the Machiavellian Intelligence hypothesis, first stated by Nicholas Humphrey and expanded and explored through field observations by Andrew Whiten and Dick Byrne, tried to relate primate mental life, social behaviour, and intelligence to humans. All this has blinkered researchers to the cognitive abilities of other species. Comparative genomics research, in which scientists look for genes that have changed in humans, compared to chimps, is the latest in a long set of such academic blinkers. Nathan Emery again:

> The argument follows thus: Primates are our closest relatives and therefore are likely to have similar cognitive abilities to humans. Anything that does not resemble human cognition must be viewed as less 'intelligent'. As primate cognition is structurally similar to human cognition, primates must be cognitively more advanced than non-primate species.

Tell that to the birds!

Crows, and related species like rooks, ravens, jackdaws, magpies, and jays, are collectively known as corvids. They seem to split humanity into friend and foe. For every distracted farmer, angrily waving his twelve-bore shotgun, there is a crow-loving aficionado, plying internet chat groups of fellow corvid enthusiasts with anecdote after anecdote about their devilish cunning, wanton enjoyment of life, sense of humour, or sheer intelligence.

Candace Savage, in her excellent book *Crows*, describes how a young female crow, on witnessing her brother's shock when an unseen petal suddenly landed by his head, picked off another, aiming it at him, to witness him jump in shock again! Was she an inveterate prankster, deriving wicked glee at his discomfort? Did her behaviour depend on getting inside her brother's mind to enable her to predict his future bafflement and annoyance? Or is Savage simply anthropomorphizing some random flit of animal behaviour?

Savage follows that up with a hilarious anecdote about a young raven who, she claimed, rolled over and played 'dead' next to a carcass he had been feeding on, the moment a flock of competitor ravens flew by. Was he signalling 'I wouldn't eat this food if I were you—it is poisoned and it has done for me!'?

How about this anecdote from a crow-lovers' internet chat group called 'crows.net'? The story concerns a rescued carrion crow who was guarding his food from the family's Siamese cats by pulling the dish away from them and marching up and down in front of it:

> The bravest of our cats decided to ignore him and eat the contents of the dish. The crow stared at her indignantly and tried to pull the dish away but the cat just crouched further over the dish. The crow stepped back and assessed the situation then went behind the cat and picked up her tail and began to pull. The cat's face was a picture!

I could go on. But why, you would be right to ask, am I recounting unprovable anecdotes in a book that is supposed to be about serious science? It is because I want to drive a coach and horses—or at least a murder of crows—through the doors of that exclusive cognitive club known as 'we humans and higher primates just like us'. And, as

we shall see, the sober field of corvid cognition research is capable
of throwing up examples of behaviour to match any uncorroborated
'tall story'.

The old *scala naturae*—the stairway to heaven by which the
early naturalists and anthropologists ordered the animal kingdom—
placed humans on the top rung of the ladder of cognition, just
below angels, with chimps and the other great apes puffing along
several rungs below. Birds, who last shared a common ancestor
with mammals some 280 million years ago, have barely hopped
onto the ladder at all! Our very name for a witless idiot is 'bird
brain'! But all that is changing. Birds—or certain species of birds—
now have scientific champions to fight their corner. Like Nathan
Emery:

> The idea that the six-layered neocortex of most mammals is the
> prerequisite for complex cognition still pervades popular cul-
> ture. Indeed, intellectually less endowed individuals in Western
> society are often called 'bird brains'. Perhaps more surprisingly,
> this view is still held by many comparative psychologists and
> neuroscientists.

One reason for this long-held, but ultimately incorrect view, says
Emery, is the confusing terminology used to name the different
regions of the avian telencephalon, or fore-brain. Traditionally,
regions in the avian cerebrum ended with the suffix '-striatum',
meaning derived from the basal ganglia, evolutionarily the oldest and
most basic part of the brain. This is involved in instinctive behaviours
such as maternal care, sexual behaviour, and feeding and has led to
the assumption that bird brains are incapable of producing flexible
or intelligent behaviour:

> It is now known that this nomenclature is based on a fallacy;
> large parts of the avian fore-brain are derived, not from
> the striatum, but from the pallium. Interestingly, the mam-
> malian neocortex is also derived from the pallium. This
> places the avian brain in a new light, where bird behaviour

may now be explained as an adaptation to solving socio-ecological problems similar to mammals, possessing hardware that is different to mammals, albeit evolved from the same structure.

If Emery is correct, then birds like corvids and parrots should bear more comparison to primates than previously thought—and they do. As he points out, Harry Jerison suggested years ago that absolute brain size may be a misleading measure of encephalization and came up with a new measure—EQ, encephalization quotient—which is a measure of relative brain size, making appropriate allowance for difference in body weight. On EQ the crow brain equals the chimpanzee brain and both are larger than one would predict from body size. Nathan Emery firmly believes that this equality of relative brain size between chimps, corvids, and parrots is mirrored in equality of cognition—in certain domains. Emery leaves no room for doubt—he calls corvids 'feathered apes'.

Emery's personal favourite anecdote is the observation of two crows at a British motorway service station, cooperating to raid the black plastic liner from a refuse bin by seizing each side of it in their beaks and lifting it up so that a third bird could help itself to the contents, before they finally spilled it onto the ground. That story is trumped by an observation from biologist Daniel Stahler, who documented collaboration between wolves and ravens. While following the Druid Peak wolf pack in the Yellowstone National Park in 1999, he noticed that the wolves were often followed by a retinue of ravens, who invariably turned up within minutes of a kill having been made. However, he also noticed instances where ravens appeared to guide wolves to prey. He watched a cloud of ravens noisily descend on the carcass of a newly dead elk calf. They were making such a din they almost seemed to be drawing others' attention to the feast, rather than trying to cover it up. Indeed, the lead wolf, some way off, noticed the shemozzle and veered through deep snow-drifts to converge on the carcass. Stahler reckons that the ravens, lacking teeth and claws to rip through the tough hide to expose the innards and meat of the elk, deliberately co-opted the wolves to do the butchery for them. Both species then warily fed together.

So much for cooperation and 'intentional' collaboration. What about tool use? In one hilarious sequence, caught on camera for one of the BBC's natural history films, crows in Sendai, Japan use moving traffic as an extended tool to crack nuts. When the traffic grinds to a halt at a certain set of traffic lights, they fly down and place nuts on the road in front of the lanes of cars. When the cars move off they inevitably run over a proportion of the nuts, cracking them. The crows, having temporarily retired to a nearby telephone wire, swoop down at the next red light in the sequence, hastily retrieve the available kernels, and replace uncracked nuts for a second try.

Another excellent book about corvids, *In The Company Of Crows And Ravens* by John Marzluff and Tony Angell, has a number of thought-provoking anecdotes about corvid intelligence. We are all familiar with talking parrots but, as Marzluff and Angell point out, corvids are a branch of the song-birds and they have a very complex social communication call system which clearly, like monkeys', refers to specific things, dangers, and social contexts. Perhaps not surprisingly they seem to be quite good mimics of other species' utterances including human speech. Their complex throat muscles allow them to get around phrases like 'I'm Jim Crow' and 'Oh my God, Oh Lord!' Konrad Lorenz's pet raven Roah was a splendid mimic and there is the extraordinary story—if true—of a hooded crow called Hansl who returned with a foot injury, having been missing for several days, and mimicked a hunter saying 'Got 'im in t' bloomin' trap!' Mickey the crow, in the National Aviary in Pittsburgh, is said to have greeted visitors with 'Hey bro, what's happenin'?' Perhaps the crow chorus singing 'When I see an elephant fly' in Walt Disney's cartoon movie *Dumbo* is not as fanciful as it seems!

So are corvids—crows, ravens, rooks, and jays—flying apes, or do scientific claims for their advanced cognition have a fictional quality—like flying elephants? Our anecdotes above have included supposed intentional deception, manipulation of the behaviour of another species, tool use, humour, cooperative tasking, and forward planning, all thought to be the exclusive cognitive preserve of humans and, though open to hearty scientific dispute, the great apes. However, a number of research groups world-wide, over the last decade, have been busy turning corvid anecdote into careful, psychological experimentation and observation and, I believe, are well on the way to a truly comparative view of the evolution of cognition. To the extent that humans, apes, and birds share the same selection pressures, evolution will sculpt similar cognitive solutions—will converge on similar ways of solving the problem—regardless of the exact nature of the cognitive substrate: the brain. It is the cognitive equivalent of the way a Mac and a PC, using different hardware and applications, can produce identical documents, art-work, and edited photographs.

The prime movers in this 'alt.cognitive psychology' movement are Nathan Emery and Nicola Clayton, and their colleagues. Clayton works principally with scrub jays and rooks. There is also formidable research from Alex Kacelnik and his team in Oxford, who work in the laboratory with New Caledonian crows; Gavin Hunt and Russell Gray in New Zealand, who design comparable experiments with the same species of crow in their native islands of New Caledonia in the Pacific; and Thomas Bugnyar, who works principally with ravens in Austria. All these researchers have designed their corvid experiments, wherever possible, to stand direct comparison with the primate research. Their experiments pose the same questions. Do corvids understand anything about the way objects in the world work—do they have any grasp of 'folk physics'? Do they have any real understanding of the way other minds work—that other individuals have desires and beliefs that may be different from their own that may leave then open to collaboration, deception, and learning—do they have 'theory of mind'? And do they understand anything about the past and the future? Can they remember lessons learned in the past, how other individuals behaved toward them, and can they use this knowledge to plan their actions in the future— do they have episodic memory? Or, as Clayton and Emery would put it, do they understand the four 'Ws'—'what, where, when, and who'? These corvid researchers explicitly address the question as to whether or not corvids can match—or exceed—the cognition of great apes in the above areas. The results are devastating for the historical 'primatocentric' view of animal cognition.

Corvids are caching birds. They store food in hidden locations— usually holes in the ground—that they dig with their beaks. There's no point in going to all the trouble of hiding valuable food if you can't remember where it is when you are later hungry, and there is no doubt that corvids have a prodigious memory of caching events, and amazing spatial memory of exactly where they cached. Clark's nutcrackers are super bird-brains when it comes to this: they can remember up to 30,000 cache sites several months after the event. How do they do it? One suggestion is that they form a snapshot of every cache location, but this seems a very inefficient method of storing such data, and one that is very vulnerable to seasonal differences that change the view—for instance, leaf-fall, vegetation growth, and snow. However, it is thought something like this is going on because, on return, they orient themselves in exactly the same position, relative to the cache site, even if they approach it from a different direction. But there is a contrasting 'landmark' theory which suggests they select a prominent feature of the cache site—a rock or a tree—and plot the location of all caches relative to that. This idea was compellingly tested by moving all objects on the right hand side of a cache site twenty centimetres to the right while leaving a large central landmark, a rock, alone. The nut-crackers displayed a recovery error

FIGURE 8. Scrub jay caching food in an ice-cube tray

of twenty centimetres when trying to retrieve caches on the right side of the arena, but were accurate to the left of the rock, which had not been altered.

Corvids are, however, by and large, intensely social birds, and so caching and memory also have a social dimension. The bane of a corvid's life is petty larceny—pilfering. Storing, or caching birds find it very difficult to go about their business away from the prying eyes of their neighbours, and there is a constant battle of wits between storers and pilferers. You would expect storers to have numerous strategies for preventing pilfering and you would expect pilferers to have large and good spatial memories for remembering the exact site of cached food, for sneaky retrieval, several days after an act of caching has been observed.

An extraordinary series of experiments in the laboratory with scrub jays, by Nicky Clayton and her colleagues, explores pilfering and confronts the question of how intentional the birds' actions are—how much they understand about their own minds, and the minds of their thieving fellows. Nicky had first noted re-caching behaviour in scrub jays over food debris left by humans, and decided to see if she could replicate the behaviour under laboratory conditions. Birds were allowed to cache wax worms from a container of

known number into ice-cube trays where all the components were
filled with sand. The storing birds were either alone, when another
bird was close by but with its view obscured, or when another bird
was allowed to observe. After three hours they were allowed to go
back to their caches and were provided with a second, sand only,
tray. The researchers therefore knew how many worms had been
eaten and how many cached originally, and now measured the number of worms recovered, the number of looks made to cache sites
(recovery accuracy), and the number re-cached. The storers both
recovered and re-cached more worms when they had been observed
than when originally caching in private. Almost all the re-caches
were in new sites unknown to the observer. It could be that the
storers predict that the observers would pinch the food given the
chance and re-cache as a deliberate counter-strategy to sow a false
belief. It could simply be that they automatically re-cache if they
have been in any way observed. To check how specific the behaviour
was they gave the storers two trays. They were allowed to cache in
tray 1 at a time when they were observed and then, ten minutes
later, they were given a second tray to cache in private. After three
hours they preferentially re-cached from the first tray. So they had
remembered when and where they had been observed. They then
added a group of birds who had previously observed and pilfered
caches from other birds. The birds were then given the opportunity
to cache and re-cache while, in turn, being observed. They re-cached
at a high rate when they had been observed—but only if they had
previous experience of pilfering themselves. Naive birds with no previous experience of pilfering did not re-cache. So previous felonious
experience seemed necessary to trigger re-caching. The researchers
reported:

> These birds appear to have transferred their pilfering experi-
> ence to the current situation, and put themselves in the perspec-
> tive of observers which may have the opportunity to pilfer the
> storers' caches in the future ... These experiments suggest the

presence of a sophisticated level of social cognition in western
scrub-jays, and one that depends on prior experience as a pilferer, but is it evidence of theory of mind?

It looks, on the face of it, as if in the world of corvids, as in humans,
you set a thief to catch a thief! The question is, as ever: How *much* do
they know?

Emery and Clayton conclude that the scrub jays display experience
projection in that only individuals with a history of larceny re-cache
when they know or think they have been observed caching. Is this a
case of mental attribution—are they 'aware' of their own previous
behaviour and does this lead to 'suspicion'? Certainly Emery and
Clayton are careful to make the point that they do not believe the
jays have a belief-desire psychology—they don't compute that the
observers will have a desire to pilfer or a belief that they know
the location of cached food. But, they say, the jays may be capable of visual perspective-taking—these claims echo earlier work
with chimps and particularly the recent work with chimps by the
Leipzig group.

Their further tests with opaque barriers present—where the storing birds are provided with two trays, one in full view and one
obscured by a barrier—closely mirror similar work with chimps,
over which Povinelli and the Leipzig group strongly disagree. When
an observer bird was present the storer preferred to cache behind the
barrier. When retrieving food they took equally from both trays, but
when re-caching they, again, elected to go behind the barrier. Were
they taking the observer bird's perspective or simply preferring a
caching site where *they* could not see the observer? In an attempt to
get round this, the Cambridge researchers did an experiment with
scrub jays in cages where one tray was put in an illuminated part of
the cage and another in a darkened area. When caching in private the
jays had no preference, but when observed they preferred to cache in
the gloom where the visual information available to the observer was

most ambiguous. When recovering food they selectively re-cached from the brightly lit tray.

So, does 'what-where-when-who' add up to theory of mind? Clayton's conclusions are notable for the care with which they are written, only leaving the door ajar to a possible interpretation of their results to mean that the jays are capable of some understanding of other individuals' knowledge states. This is how she puts it:

> Our findings suggest that scrub jays remember who was present during earlier caching events. This ability need not be based on a human-like episodic memory, because these 'what-where-when' memories may exist without the jays mentally reconstructing the past. Similarly, keeping track of 'who was watching when' need not require a human-like 'theory of mind' in terms of attributing unobservable mental states ... Although the ability of nonhuman animals to reason about another's mind continues to elude definitive study, our study provides evidence to suggest that a non-human animal might discriminate between individuals with different knowledge states.

Contrast this, however, with much more ebullient prose quoted by Clayton on her page in the University of Cambridge website. Clayton's and Emery's report in *Nature* of possible mental time travel in scrub jays had drawn this response from Prof. John Pearce, of the University of Cardiff:

> The demonstration of Clayton and Emery that scrub jays are more likely to move food from one hiding place to another, if they were observed by other scrub jays when originally hiding the food, by itself is a remarkable finding. But to demonstrate that this effect depends upon the birds already having stolen food hidden by other scrub jays is quite extraordinary and has far-reaching theoretical implications. It suggests that scrub jays may possess sophisticated thought processes that allow them to anticipate and outwit the actions of other birds. If this is true,

then scrub jays will be the only non-human species that can be
said to possess a theory of mind.

It is quite clear where Clayton's and Emery's hearts and minds are.
They veer toward a higher-order mentalistic interpretation of their
results. But they are being cautious and letting others blow their
trumpets for them!

These results have received substantial backing from the work of
Thomas Bugnyar and his colleagues in Austria, working with ravens.
Bugnyar's observations on deception were not planned. His original
experiment was to investigate social learning and scrounging among
young ravens over concealed food. However, a game developed out
of thin air between a subordinate and a dominant male, Hugin and
Munin. (In Norse mythology the two ravens, Hugin and Munin, were
the eyes and ears of Odin. They would fly through the nine worlds
and bring him news of what was going on. Munin means 'memory'.)
The food—pieces of cheese—was concealed in film tubes with lids.
After original training on the nature of the task, several clusters of
film-tubes, some baited and some not, were spread over the outdoor
foraging area. The clusters of film tubes were colour-coded, only
tubes of one colour were baited on any one day of the trial, and
the colour-coding of the tubes containing the cheese was changed
every day.

The subordinate male, Hugin, proved to be a star. He uncovered
two thirds of all baited tubes, was nearly always the first to discover the baited clusters, and very quickly learned to avoid clusters
that proved, on inspection, to be unrewarding. Munin, the dominant
male, spotted Hugin's conspicuous success and turned scrounger,
pouncing on Hugin and managing to filch food from him on over
80% of the occasions when Hugin retrieved it—a very high cost to
Hugin. Hugin began to alter his behaviour and moved occasionally to
unrewarded clusters. Munin followed him, expecting to profit, but at
that point, Hugin would leave Munin poking away at the decoy, dash

back to the rewarded and not yet fully exploited cluster, nab the food, and make off. The whole sequence of events took less than fifteen seconds. As if aware of the 'cry wolf' principle, Hugin moderated his 'leave for unrewarded cluster, return to rewarded cluster' subterfuge, but he would do it at least once a day, his moves increasing with the frequency with which Munin filched from him. When Munin hesitated before following him to an unrewarded decoy cluster, Hugin persevered at his play-acting longer, pecking at more unrewarded boxes, which suggested to Bugnyar that he recognized the relationship between what he was doing and Munin's reaction. So he *appeared* to be actively misleading Munin, which puts his behaviour on a par with the intentional deception claimed for wild chimps by many primatologists.

Bugnyar, however, did not jump to the conclusion that Hugin had got inside Munin's mind—that he was messing with Munin's beliefs. He sat on the fence, suggesting that a more 'parsimonious' conclusion could be that Hugin was manipulating behaviour rather than mental states.

Bugnyar then put pilfering and gaze-following together in a set of experiments with his ravens that are entirely complimentary to those of Clayton and Emery with scrub jays and Hare with chimps. In the competition between cachers and pilferers, would the caching bird's behaviour, when it retrieved cached food, depend upon whether it was observed caching and whether or not obstacles interrupted the view of the observing bird?

In the first experiment a raven was faced with the choice between retrieving its cache in private, in the presence of a subordinate who had observed the original caching from an adjacent compartment, or when a subordinate had not observed the original caching because a curtain had obscured its view. The prediction was that the storer would be more likely to retrieve and re-cache its food if the original caching had been done in the presence of an observing bird, than if the potential pilferer could not see, or the storing bird had been alone—and this is exactly what they found.

Much has been made—far too much to my mind—of the tool-using capabilities of chimpanzees in the wild. Do these behaviours prove that chimps possess genuine insight and innovation? Unfortunately, there is little or no evidence or observation of that 'eureka!' moment of pure insight which heralds novel tool use. Povinelli's work on folk physics with chimps suggests they simply do not have enough deep understanding of the unobservable properties of the physical world to have such flashes of insight. Chimps have been nut-cracking and termite-dipping for millennia and no one can have any idea how the practice first came about. According to one of Aesop's fables, say Emery and Clayton, a corvid Archimedes placed a number of stones in a pitcher of water until it had displaced so much that the water level came within reach of its beak. Eureka! Do corvids in the real world show such insight? As we shall see, fact is almost as strange as fiction, as far as corvids are concerned.

The classic experiment that suggests the answer is 'yes' concerns Berndt Heinrich's work with hand-reared ravens. Here, pieces of meat were suspended from a branch by a string. The only way to obtain the meat was to pull the food up, and, indeed, none of the ravens attempted to get at the meat by pecking at it in flight or by hovering around it. To get the meat, the ravens would have to pull up about five relays of string with foot or beak. Three of the ravens opted for this solution at first go. If presented with two strings, one with meat attached, one with a small rock, they either immediately selected the meat or immediately switched to the meat after one yank on the rock. Crossing the strings did fool most of them, but one bird sorted it out and selected the correct string. Substituting one string for a coloured lace resulted in the lace being pulled only if it had meat attached, suggesting that the birds had not merely created a rigid association between a type of string and food; and, finally, when one string held meat and the other a sheep's head, the birds clearly avoided the option which would have fed them better but was too heavy to pull up. Heinrich concluded that they were using insight because most behaviour showed itself immediately. If they had rehearsed the behaviour at all it could only have been by mental simulation in their heads.

What about tool use? The New Zealand duo of Russell Gray and Gavin Hunt, and their various colleagues, have compiled compelling, vivid, and detailed observations of tool use in the wild by New Caledonian crows on the several islands of the New Caledonia archipelago in the south Pacific. Here they look at the manufacture of both twig and Pandanus leaf tools. Pandanus is a tropical and semi-tropical shrub with long, trouserbelt-wide, tough leaves. To imagine the Pandanus leaf tools you need the image of a saw-blade in your mind. The crows cut the 'saw' out of the leaf using a series of stepped nips and rips. There are three basic variations in design—wide, narrow, and stepped tools. The researchers can always get information on this range of design because every leaf tool leaves its 'negative' in what remains of the leaf-blade on the plant. They observed that the birds have a bias for making tools on the left edges of the Pandanus

FIGURE 9. New Caledonian crow cutting a leaf tool out of a Pandanus leaf: (a) The crow is making a cut to form the tapered end of a tool, working toward the leaf tip while standing on the left edge of the leaf. (b) The crow is making the wide end of the tool and working back toward the trunk.

leaf blade, with very faithful manufacture of the step design so each tool of its type is identical.

To make the most complex, stepped, tool, for instance, the birds begin by fashioning the narrow end of the tool, starting out nearest the base of the leaf and worked outwards, cutting across fibres and ripping along the plane of the leaf. They then jump toward the distal end of the leaf, making two deep cuts across fibres to cut out the wide end, before working back toward the base in a long rip to join up with their earlier work. Skill was needed to make these rips along the plane of the leaf converge so that the blade of the tool could be neatly severed from the parent leaf. The tools were made by a one-step process. No further cutting and shaping was done once they had been removed from the plant. Since all three leaf-tool designs occur within the same geographical area, the researchers conclude that tool design has evolved cumulatively from simple to complex and they have circumstantial evidence that the technique is transmitted socially to successive generations of younger crows.

This very distinctive method of highly reproducible tool manufacture contrasts favourably with tool manufacture in primates, say Hunt and Gray. Although chimps will strip twigs and shorten them to length they show little understanding of the idea of hooks, and although orang-utans make tools to snare other branches, which has been described as 'hooking', Hunt and Gray consider it more like 'branch hauling' because there is little evidence for obvious and literal hook manufacture.

The crows use these barbed, stepped tools to extract grubs from holes and nooks and crannies in vegetation. Because Pandanus barbs point forwards, when on the plant, the crows hold the tool by the wide end, barbs facing away from them, and raddle out the grubs with the barbs. If the stepped tools are not rigid enough to lever food out of holes the barbed edge is forced, or brushed, up against the larva to attach it by the barbs and then pull it out. If tools are

FIGURE 10. Selection of stepped Pandanus leaf tools

unsuccessful they are rapidly thrown away and new tools made. Hunt and Gray think the crows select the correct size of tool to use by measuring the distance to the food. At one site they noticed that the crows not only cut tools from the left-hand edge of Pandanus leaves but held the tool with the non-working end to the left of their heads. This, they say, is the first reported case of lateralization in both the use and manufacture of tools in non-humans.

Now, playing the curmudgeon, you would have to say that all this could be done using procedural rules or rote behaviour, but Hunt and Gray would have us see it otherwise by claiming that, unlike chimpanzees, crows grasp some of the functional properties of their tools. The elaborate cutting and ripping manufacture of Pandanus tools, they argue, suggests the end result owes more to

Break points

Break points

FIGURE 11. Manufacture of hooked twig tool by New Caledonian crows

manufacturing technique than the inherent structure of the raw material and its constraints. This is a very important point because chimps have never been seen to extensively adapt the raw materials they work with.

Their observations of twig hook manufacture by the same crows add weight to the argument. The birds first select a promising fork in a twig. They then break off one arm of the fork just above the junction and discard it. This forms the hook. Next, they break off the remaining twig just below the fork—so that the hook is left as the terminal end of the twig—this is the twig tool. They then nip the other arm of the fork at longer length to use as the shaft and then fine sculpt the hook of the tool to remove small pieces of wood from the hook and sharpen it. They also remove leaves from the shaft of the tool. Hunt has watched one crow and its dependent juvenile make a ten-piece hooked-tool set by this relatively invariant

FIGURE 12. New Caledonian crows using hook tools in the wild

three-step process and claims it has four features previously held to
be exclusive to hominids: a high degree of standardization, the use of
hooks, handedness, and cumulative changes in tool design. Although
this tool manufacture betters anything observed with chimpanzees
the jury is still out on whether or not it provides robust evidence for
'folk physics'. There is no evidence, for instance, that mature crows
'teach' juvenile crows how to do it. However, twig-tool manufacture

is very impressive because there is no natural hook—it has to be perceived in and fashioned out of the complex geometry of natural vegetation.

While Gavin Hunt and Russell Gray have been investigating insight in the manufacture of tools among New Caledonian crows in the wild, Professor Alex Kacelnik and his colleagues have been devising a series of extraordinary experiments with New Caledonian crows in the laboratory in the Zoology Department at the University of Oxford. New Caledonian crows are slightly smaller and sleeker than the crows we are used to in Europe and North America. Their plumage is very shiny—with that almost iridescent quality you see in oil or petrol in sunlight. They are also very naughty! When I visited the experimental area with Alex he told me to put on a special hat with ear-flaps—a little like a 'Biggles' flying helmet—'for your protection'. I soon found out why. As we entered the enclosure, the crows greeted me by flying around in a circuit, fleetingly alighting on my head at every turn, as if they were light aircraft doing 'circles and bumps' at an airfield! One of them alighted by my foot, shot me a wicked glance, and proceeded to untie my shoe-lace. I resisted the urge to say 'If you're as darned clever as Alex says you are, let's see you tie it back up again!'.

In a set of experiments that started in 2000, Kacelnik's group set out to test the crows' ability to select tools appropriate to a task and to rate a tool's length or diameter for appropriateness in reaching food. If they could work it out without undue training, it would be indicative, at least, that they were using some degree of insight. The experiments were constructed to bear direct comparison with techniques used by Visalberghi and Povinelli, and described for chimpanzees in chapter 8.

For the first experiment, food was placed over a range of distances within a perspex tube, closed at one end but with a central hole bored through the plug at the other. They provided the crows—a male and a female—with a tool box of ten bamboo skewers, cut to a range of lengths, and held upright in a row of holes drilled into a flat piece of wood, rather like a box set of drill bits. This was turned round between trials to make the shorter skewers closer to the tube on some trials, and the opposite true in others.

The birds always inspected the meat in the tube, through the wall of the tube, and through the open end of the tube, before selecting a stick. Sometimes a bird would grab a stick, discard it, and select another, sometimes they held the stick by one end so that it was in line with beak and head, sometimes they held it some way down

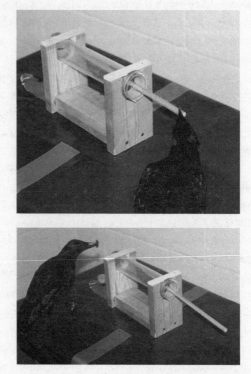

FIGURE 13. Betty selecting a tool of appropriate length to poke food out of the tube

its shaft with the remaining part of the stick lying against the side of the head. This gave them a little more control over the manipulation of the stick. The birds chose the longest tool on half the trials, and the matching tool for distance on a quarter of the trials. They seem to have been using a rule like 'either choose a stick closely matching the distance to the food, or choose the longest tool in the set'. Both strategies resulted in the crows getting the food (which was surely the point for them rather than being pernickety and matching for length as if to please the experimental design of the researchers!)

What would have happened had they been required to find their own sticks? A second experiment tried to test this by moving the tool block across the room and placing a barrier between it and the tube. The female never did retrieve any sticks for use on the tube, though she did remove some of the sticks and use them elsewhere. The male did much better but he took much longer than in the first experiment. This was due to a lag in deciding to do anything at all. Once he made his move he retrieved the food just as fast as before. Typically he would sit on his perch for a long time, then he would fly to inspect the food, as before, then fly over to select a stick. On two occasions he selected a stick that was strictly some two centimetres too short but succeeded by using the 'tip' method of holding the stick by the tip of his beak, thereby making up for the short-fall in stick length. On the other two occasions he selected a stick that was far too short, returned very rapidly to the tool box, selected a longer stick, and successfully retrieved the food.

In a second set of experiments, Kacelnik's team tested to see if the crows were capable of selecting a tool diameter appropriate for the job. They used an L-shaped tube with the horizontal arm capped with bungs with holes bored through them. The other arm of the tube pointed downwards and the bird had to select a stick that could be inserted through the bung hole to push a cup containing the food 'down the well'. They were given sticks of three different thicknesses—thin, medium, and thick—and the sticks were either presented as one loose stick, with the other two bundled together, or all three bundled.

The female crow always immediately took the bundle of sticks off to a nearby perch, tore off the wrapping, and selected one stick—invariably the thinnest. When all three sticks were loose, she still chose the thinnest stick on all 24 occasions. The thin stick was appropriate on all occasions, and would certainly have been the lightest, and perhaps the easiest to hold in the beak. The crow still preferred to fly off with the bundle and dismantle it if the thinnest stick was in it, even though a thicker, but still serviceable stick was the one lying loose.

In the second, related experiment, the tube was as before, but instead of a bundle of sticks, a bushy oak branch was placed near the tube. This time a pair of crows was used and they managed to retrieve food in all but one of thirty trials, and always within twenty minutes. After inspecting the tube, they would fly to the branch, strip the leaves off an appropriate twig, and nip it off.

Because of the female crow's overwhelming preference for the thinnest rod, the first experiment could provide no precise evidence that she was taking the diameter of the hole into account. However, in the second experiment, tools were quickly rejected if too large and there was evidence that, as the hole got wider, the twig selected got wider. In all cases it was the 'working end' of the tool which matched the aperture—an impressive feat given the irregular shape of most twigs. This really does show some appreciation of the task. Given these apparent skills, why did the crows select, on six occasions, a twig that was too short to topple the cup down the tube? It could be that pushing is more 'foreign' to crows than pulling: holes in trees are blindly ending—there's only one way out! Or maybe they were guesstimating the distance from the end of the tube to the cup, and not the full travel to the downward spout? Or it may be down to technique—with the crows sometimes using the

FIGURE 14. Betty

snooker cue technique of giving the cup a sharp knock to send it flying. Again, it was the selection and modification of tool material—clearly related to having a precise end function in mind—that is the most impressive feature of this test. This is a rare and important finding.

Kacelnik's prize performer over the years has been a female crow—Betty. Betty first came to fame thanks to reports on early experiments in Kacelnik's laboratory involving straight wires and wires bent into hooks, to see if the crows would select the correct tool to lift a small container of food—usually a piece of pig's heart or a waxmoth—out of a length of perspex tube. This is directly analogous to pulling a pail of water up the shaft of a well. Betty had selected a hook-shaped piece of wire and was just about to fish with it when the dominant male, Abel, tore down and snatched it from her. To the surprise of the researchers, Betty calmly selected a straight piece of

FIGURE 15. Betty using a hooked tool to retrieve a bucket of food from a well

wire, inserted the end of it into the mesh of the cage, and bent it with her beak into a hook, before proceeding to use it and obtain food. Sure that it was a fluke, the researchers repeated the experiment at least ten times, provisioning Betty only with straight pieces of wire. In nine out of ten occasions she repeated the manufacture, immediately and spontaneously.

Kacelnik's team went on to investigate further the extent to which Betty (and Abel) understood the folk physics of hooks. They claim, quite rightly, that humans and crows are the only animals investigated so far that make tools with a precisely determined final shape. This is distinct from chimp tool manufacture where, although the technique is repeated, the end product is not determined in detail by the manufacturer—it is often inherent in the nature of the material. With hooked twig tools, stepped Pandanus leaves, and bent wire strips, in contrast, shape and function are imposed onto an unstructured substrate.

The experiment the team conceived was meant to compare specifically with Povinelli's apparatus, as discussed in chapter 8, where straight or curved piping was given to apes to see if they could re-shape it to poke through a hole and retrieve an apple. You will remember that their performance was dismal. Betty was subjected to three experiments with a novel material—aluminium strips—which required her to bend or straighten the strips in order to obtain food. She straightened the strips appropriately on four out of five occasions, and used novel techniques for bending the material. However, the team stop short of a belief that she really understood causality. Sometimes she probed with unmodified tools before changing them, and, occasionally, she modified her tool but fished for food with the wrong end. Nevertheless, her performance was impressive. The team concluded:

> Gauging New Caledonian crows' level of understanding is not yet possible, but the observed behaviour is consistent with a partial understanding of physical tasks at a level that exceeds that previously attained by any other non-human subject, including apes.

Unfortunately, at this point in her stellar career Betty died. But before she expired, in the manner of a true star, she saved her best for last. Betty's swan-song was a very complicated test comprising one horizontal tube with a piece of food deep inside it and a very short stick beside it, far too short to reach the food. There were four other identical tubes, spread out in a fan nearby, one of which had in it a stick long enough to reach the food, but this stick was pushed way back into the tube, so that it could only be got out using an intermediate tool. The short tool would not retrieve the long one. In three other tubes were intermediate-sized sticks. The task was to see if Betty could use the short stick to get out the intermediate stick to get out the long stick to get out the food. If Betty had been learning all along through reinforcement—an immediate reward for choosing the right implement—she would not be able to do this test because the positive reinforcement only came at the end when the food was retrieved. The intermediate steps bore no immediate reward. Betty came in, tried momentarily to fish for the food with the short stick, immediately whirled around, took a quick dab at the long stick with the short stick, realized it would not work, hopped over to one of the tubes containing an intermediate stick, fished it out with the short stick, immediately threw the short stick away, retrieved the long stick, threw the intermediate stick away, and went straight to the food and retrieved it with ease! All in a matter of seconds and on the first trial!

The New Zealand research group have also been investigating this meta-tool use (the intermediate use of a tool to modify another tool before its use on a target) with crows. They deliberately set out to adapt an experimental design that had been used earlier with apes, modifying it for crow use. Pieces of meat were placed in 15-centimetre-deep horizontal holes, drilled into a tree trunk, situated about two metres away from two identical 'tool-boxes'. The front of each tool-box was covered with vertical bars that allowed the crow to insert its beak but not the entire head. They placed an eighteen centimetre stick tool four centimetres inside one toolbox. This was too far in to be fished out using the beak, but, once out, would be more than adequate to reach the meat. In the other tool-box a stone was placed in a similar position. In front of the tool-boxes they placed a five-centimetre stick. Three of the crows, Icarus, Luigi, and Ruby, spontaneously used the short stick to retrieve the long stick and then went on to successfully retrieve meat using the long stick. Other crows also solved the meta-tool problem, given more time, or solved the meta-tool problem but messed up because they then took the short tool to the hole. Ruby, Joker, Luigi, and Colin occasionally attempted to use the short or non-functional stick, but only when they occasionally failed to obtain the long stick with it—leading to optimistic failure on the overall task. These results are the equal of any data from apes.

Had the crows genuinely understood that they first had to take a short tool to get the long tool to reach deeply embedded food because only that could achieve physical contact between stick and food such that one could move the other? In a second experiment they reversed the position of the tools in the tool-box. Now the long stick was outside and the short stick (now utterly redundant) was inside. Would the crows fall for it? All six crows initially fished inside the tool-box using the longer tool. However they only did this in the first block of trials and most of the crows abandoned it in a flash, taking the long stick directly to the hole.

Hunt and Gray believe they have demonstrated some very advanced cognition. Making or accessing tools to make tools to get food requires that the crows delay the temptation to go for immediate gratification by trying to go straight for the food. It also requires that they show hierarchically organized behaviour so that they can integrate novel behaviours (tool→tool) as a sub-goal of the main goal of tool→food. It is this crucial meta-step, it is widely assumed, that allowed human ancestors to progress from simple percussive techniques to the knapping of flint cutting-tools. Work by Povinelli's group, and the Leipzig group, among others, have examined chimpanzees' abilities to use short sticks to retrieve long sticks which can then be used to fish for food and it is clear that, in these specific areas of cognition, crows can be a match for chimps.

Back at the University of Cambridge, Amanda Seed has also been picking up Povinelli's gauntlet and asking, of rooks, 'Do they understand anything about the physical principles and causal regularities that underlie their tool use? In other words, do they really have a folk physics as we do?' To find out she used a number of complex variations of the trap-tube test already much used with monkeys and apes. Here the apparatus is a perspex tube with the food placed between the vertical drop of two traps that can be made functional or non-functional. The rooks had to discriminate between the two and push the food appropriately, from one end or the other, so that they could retrieve it and not lose it down the trap. One bird, Guillem, proved to be an absolute star—passing the most difficult test within the first block of ten trials, scoring nine out of ten and ten out of ten respectively. Although the majority of the rooks failed to match up to Guillem's Olympian performance they performed well on a version of the task every bit the equivalent of the hardest of this type ever passed by a primate, say Seed, Clayton, and Emery.

The New Zealand group have taken this paradigm one step further. Three out of six crows laboriously managed to solve the trap-tube test over 100 trials and were then transferred onto a corvid version of the trap-table test used by Povinelli and others for apes. They solved it on the first trial, suggesting to these researchers that they had transferred some level of causal reasoning, however weak, to the new problem rather than solving it either by chance, learned visual cues, or by generalizing the tactile nature of the task. Other scientists—partly because it took so long for the penny to drop in the original trap-tube trials—are less convinced! However, this result, the New Zealanders point out, is in stark contrast to a recent study where four ape species were unable to successfully transfer from trap-tube to trap-table.

Nicky Clayton has already demonstrated some degree of mental time-travel—or episodic memory—in her scrub jays. They clearly remember when they cached food, whether they were overlooked, and who was watching them at the time. They understand the 'when, what, and where' of caching and pilfering. In tests using wax worms and crickets, which degrade at different rates, plus less preferred peanuts, which have a very long cache life, they rapidly learned about the different degradation times and switched, after four hours, from digging up wax worms in favour of digging up peanuts. When the test was made more complicated by using two insect cache materials—mealworms, which tend to decay after about twenty-eight hours, crickets, which last for up to five days, and long-life peanuts—they appeared to track the length of time each type of cached food had been in the ground and switched their retrieval preferences accordingly, not bothering to retrieve cached food that was past its sell-by date. Clayton has very recently devised a wonderful experiment to test time-travel in scrub jays still further. Can they use their knowledge of the present to plan for the future—a mental property thought uniquely human?

The jays were put into an open chamber with food freely available, but in powdered form, so, although they could eat it, they couldn't cache it. In the evening, on some nights they were shut up in one bedroom and on some nights they were shut in another bedroom at the opposite end of this common area. Over six nights they were allocated their bedroom on a random basis. Every morning when they woke up they were hungry and in one room breakfast was served but the other had no room service—the unfortunate inmates had to go hungry. All of a sudden, the evening before the main trial, they were given plenty of food in whole, as opposed to powdered, form. Over 90% of the birds cached their food in the room where experience had taught them that breakfast would not be served the following morning! They seemed to be planning for the future because they were doing this when there was lots of food freely available in the common area and they had eaten their fill. It was as though they were imagining a tomorrow in which they would be hungry, not satiated, and where there was a high chance that they would wake up in a dormitory run by a scrooge! Clayton even staged a follow-up experiment where two types of evening food were served, but only one type was made available for breakfast the following morning.

The jays only cached the food type that experience had taught them would not be on the breakfast menu!

There will always be argument as to whether corvids have demonstrated true episodic memory, or true causal reasoning, as we humans exercise it, and whether or not they have demonstrated equality with or supremacy over apes in these specific cognitive areas. So are corvids 'feathered apes', or not? Alex Kacelnik does not like Nathan Emery's moniker because he feels it implies across-the-board cognitive equality, and he does not think that this is the case, or, at least, that it has been proved. But he agrees that the Cambridge group have done astonishing work on three aspects of cognition which challenge primate supremacy: the idea that jays can project their own past experience of pilfering onto observers and use it to decide whether or not to re-cache food; the idea that jays can plan for the future and provision certain dormitories with certain food in a time when they will be in a different state (hungry) to the one in which they make the decision to cache (satiated); and the idea that rooks really do understand something about physical causality when retrieving food in the trap-tube. To this I would add Kacelnik's work, also on trap-tubes, of the manufacture of novel tools by bending or straightening aluminium strips, and the employment of intermediate tools to access other tools to recover food—bolstered by work on 'meta-tools' by the group from New Zealand.

Does this prove that corvids have 'theory of mind' and are therefore cognitively closer to us than chimpanzees? Of course not. None of the experiments I have described can satisfy us that corvids have human-like cognition that makes them aware of the unobservable mental life of other individuals, or the unobservable phenomena that govern the way the physical world behaves. Equally, they cannot prove that corvids do *not* have some awareness of these things. Danny Povinelli has recently—and, in my opinion, fairly—extended his curmudgeonly approach on chimpanzee cognition to the experiments on corvids and it may be that all comparative cognitive psychology is in limbo until new experimental paradigms arise that all can agree will resolve the matter.

But my argument does not, thankfully, depend on proving whether or not corvids think like humans. It does depend, however, on showing that, in two species that parted company 280 million years ago, performance is either very similar, or corvids might even have the edge. Bird brains, in specific contexts, are a match for chimp brains. Here we have a class of animals—birds—which until recently were regarded as little more than simple robots, more than holding their own with a species of primate known to be very closely related to us. And, although I have neglected them in this book because I think that with dogs and corvids my point is made, I could have presented data from goats, elephants, sea-lions, and cetaceans that would have further eroded this unduly chimp-ist view of cognition, where apes, because of their genetic and taxonomic proximity to us, are assumed to share cognitive structures with us that the rest of the animal kingdom lacks.

The important take-home point is that cognition is a tool to do an adaptive job, and when social and ecological problems are similar it can be expected to solve them in similar fashion, whatever the species. Claims for chimpanzee tool use, deception, manipulation of others, and insight can no longer reinforce claims for their evolution-ary and genetic proximity to us, but only show that, like big-brained corvids, they have shared some of the same social and ecological problems as us. Any species that does so will evolve the necessary, and *functionally* analogous, cognitive structures to deal with them. The argument by analogy is undone.

The psychologist Sue Savage-Rumbaugh, famous for her work over the years with Kanzi the bonobo, wrote a book about her research experiences—*Kanzi: The Ape at the Brink of the Human Mind*. But I see Kanzi, and the rest of the 'prodigy' great apes that adorn the popular literature on the subject, as doing little more than pok-ing around in the foothills of the mountain that is the human mind. If, however, you are happy with the very arguable title of Rumbaugh's book, you must not be upset if, in a few years time, when you run your eye over the popular science titles in your favourite bookshop, you come across *Corvids: Perched on the Lower Branches of Human Cognition*, or *Dogs: Barking Up the Right Tree of Human Intelligence*.

CHAPTER 10

Inside The Brain—The Devil is in the Detail

Imagine you are standing with both your arms held out in front of you and slightly to the side, palms upward—a little like a human version of the scales of justice that adorn the roof of the Old Bailey law courts in London. On your left palm sits a human brain—careful now, it is heavy: at over 1300 grams it weighs as much as a litre bottle of wine. On your right palm sits the brain of a chimpanzee. At just over 400 grams it is less than a third the weight. If an anatomist had placed a chimpanzee's tongue on your palm, instead of its brain, it would have weighed more! Now think for a moment about their owners. The hominid which owns the brain on your left palm built the scales of justice I have asked you to pretend to be. And the famous London law court on which it sits. And the concepts of justice, moral judgment, human rights, and the due processes of law taking place inside it. The hominid which owns the brain in your right palm has never strayed from several carefully circumscribed areas of African tropical forest in all the 6 million years of its existence. It couldn't pretend to be the scales of justice because it is not capable of metaphor. It has, as far as we know, no understanding of the mental states of belief

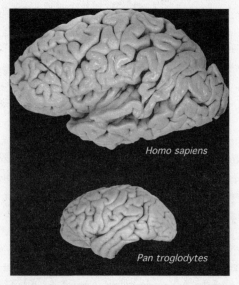

Homo sapiens

Pan troglodytes

FIGURE 16. Comparison of chimpanzee and human brains

or desire, and lacks anything but the most crude foundation for social tolerance and justice.

Let us take our brains into the laboratory and look at them in a little more detail. The outer surface of the human brain is highly convoluted with folds, or gyri, separated by fissures. This is the cerebral cortex—in fact most of it is termed neocortex—a structure which is unique to mammals. What about the chimpanzee brain? Although much smaller, it too is highly convoluted—though perhaps not quite so much as the human brain.

Now for a little butchery. Lay the brains on a slab, take a sharp knife and cut both brains straight across from left to right to reveal their cross-sections. To the untrained eye there is little or nothing, other than size, to suggest any gross anatomical difference between the brain of the human and the brain of the chimpanzee. The chimp brain seems to be a 'mini-me'. One would think that something more than just an increase in size was responsible for some of the obvious cognitive differences between us and chimps. Size cannot be everything. Yet, for over 150 years, biology has brushed aside any suggestion that the cognitive differences between us and our primate cousins could be due to anything more than scaling up. The phenomenon is known as allometry. As you proceed through the primates, from the tiny prosimians at one end of the scale to us humans at the other, brain size gets bigger. But allometrists insist that, because of developmental constraints, brains increase in size in such a way that the different parts of the brain—frontal, temporal, and parietal cortices, cerebellum, etc.—retain the same proportionate increase: no one part of the brain increases disproportionately in size at the expense of another part. The human brain, therefore, is just a big ape brain and, if there are differences in cognition, they are attributed to larger amounts of something, cortical neurons perhaps, rather than departures from allometry. Two staunch allometrists, Barbara Finlay and Richard Darlington, claimed, back in 1995, that the first principal component—brain size—explained more than 96% of the variance in 11 major brain structures, including the cerebellum and neocortex. 'This led to the conclusion', says Jim Rilling, a neuroscientist at Yerkes Primate Centre, 'that intractable developmental programmes force mammalian brain growth to follow predictable allometric trends in which individual brain structures enlarge mainly by concerted enlargement of the entire brain. In other words, all mammalian brains are basically smaller or larger versions of the same plan.' Evolutionists find this idea of brain allometry very dull indeed. They thrive on variation, it is their bread and butter, and they are forever looking for differences—specializations—adaptations—rather than similarities. To neuroscientist Todd Preuss, a colleague of Jim Rilling at the Yerkes Primate Centre, this 'allometric bauplan' simply made no sense at all.

We have recently discovered, says Preuss, that there are hundreds of genetic differences between the brains of humans and the rest of the primates, but the anatomical and functional differences that should correspond to them are woefully lacking. The reason lies in the history of how neuroscience research is done. Since the 1970s, neuroscientists have been able to trace detailed pathways of neurons throughout monkey brains because they can open them up, inject fluorescent dyes, and stick electrodes into them—usually sacrificing the monkey in the process. So complex wiring diagrams for things like macaque brains have long existed. But you can't do comparative work like this in humans. Therefore the vast majority of neuroscience is done on animal models—monkeys and rats. Human brain structure and function is extrapolated from these animal models. It leaves us, therefore, Preuss flatly states, without a neurobiological theory of human nature:

> While the strangeness of humans may be commonplace to anthropologists it is not an idea that neuroscientists typically find easy to wrap their minds around. This model-animal approach to human nature has important limitations. It fosters the view that humans, as well as the animals we use as proxies for understanding humans, are somehow typical, and thus dignifies only the study of ways in which animals resemble one another. In the agenda of modern neuroscience, the evolutionary specializations of the human brain scarcely seem to rate a footnote. But if we dismiss human brain specializations as mere variations, we trivialize the very features of the brain that make us humans, rather than monkeys or mice or fruit flies. Evolutionary specializations are not trivia—they are the essence of human nature.

Allometry is a Mekon theory of human cognition, after the melon-headed Mekon from Mekonta, Dan Dare's super-intelligent adversary in the sci-fi comics of the 1950s. But Preuss, and other evolution-minded neuroscientists, argue that it is time to ditch this 'bigger, better but similar' notion of brain evolution and concentrate on the task of finding out whether or not advanced and perhaps unique human cognition arises out of unique anatomical specializations in the human brain. Neuroscience is only just beginning to make inroads on this daunting task. As Todd Preuss says: 'What slim pickings! It seems extraordinary that neuroscience has so little to offer on a matter so fundamental as what it is about our brains that makes us human!' Finding Todd Preuss's 'slim pickings' has not been easy. There is only one genuine novel structure found, to date, in human brains, and we have to thank Preuss himself for the discovery that certain cell layers in primary visual cortex are more complex in humans than in either macaques or chimpanzees and may have contributed to our better ability to interpret the visual field in 3-D, perhaps allowing us greater manual dexterity. As a general rule, humans and chimps share all the main brain areas, and what differences there are seem to lie more in relative size differences and new functions running on old structures. As with most of life, the devil is in the detail and comparison of human and chimpanzee brains becomes more interesting the deeper one looks.

Let us start with the most obvious difference between human and chimpanzee brains—their size. No one disputes that the human brain is about four times larger than you would expect for a typical anthropoid primate of our body size. But what about the most evolutionarily recent addition to the brain, the cerebral cortex, and its four principal components—the frontal lobe, sitting behind our forehead and extending about half way back; the temporal lobe, sitting on the side of the brain forward of the ears; the parietal lobe, sitting on the top of the brain toward the rear; and the occipital lobe (which contains the visual cortex), which sits at the rear of the brain? Even at this gross level of anatomy there is scientific disagreement as to whether or not there are any real relative differences in humans.

Take the frontal lobes, for instance. Here is one area of the brain which has been considered for years to have expanded disproportionately in humans, in comparison both with the rest of the human brain, and with the frontal cortex of all the great apes. Indeed, the frontal lobes, with their capacity for rational thought and detailed forward planning, seem to get closest to the very idea of what it is

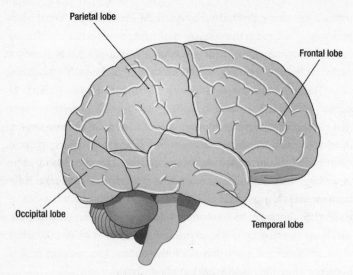

FIGURE 17. The four lobes of the human cerebral cortex

to be human. Yet, when Katerina Semendeferi and Hannah Damasio compared fMRI brain-scan data for humans and chimps, in the late nineties, they concluded that although the human frontal lobe is bigger in absolute size than that of a chimp, it is not relatively bigger—humans do not have a larger frontal lobe than expected for an ape brain of human size. In other words, if you took a chimpanzee brain and quadrupled its size the frontal lobes would be in the same proportion to the rest of the cortex as in humans—just over a third.

Yet Jim Rilling, the scientist who actually took the fMRI scans Semendeferi used for her analysis, interprets them differently. He maintains that the cerebral cortex as a whole (which of course includes the disputed frontal lobes) *is* significantly larger than expected—it has increased disproportionately. 'In other words', he says, 'the human brain is not just an enlarged non-human primate brain, it is a different brain; one dominated by cerebral cortex.' Rilling backs that up by referring to another data-set of post-mortem brains where the human neocortex *is* disproportionately large compared to the rest of the brain. So what is really going on? Rilling maintains that the bulk of the disproportionate increase in human cortex is due to expansion of the temporal lobe. And much of that is due to an increase in white matter which consists of axons—the long-distance 'cabling' of the brain—that link the temporal and frontal cortices. Rilling thinks that most of this relates to the emergence of language because it has become clear over the last few years that large portions of the lateral surface of the left frontal and temporal lobes are involved in language in addition to the more prosaic language areas we already know about—Broca's and Wernicke's areas.

Wernicke's area lies toward the back of the left side of the brain, near the junction between the temporal and parietal cortices. It is part of a bi-lateral structure called the planum temporale, which is actually part of the auditory association cortex and processes sounds that come in from the ear, before sending information to other parts of the brain. Wernicke's area is known to be involved in language comprehension—particularly the meaning of spoken words. Since language is heavily lateralized to the left side of the brain it will come as no surprise to find that most studies show the planum temporale in the majority of human subjects to be much larger on the left than the right. For a long time this was thought to be a uniquely human feature, but, in the late 1990s, Patrick Gannon and Ralph Holloway investigated the planum temporale using post-mortem tissue of human and chimp brains and found, to everyone's surprise, that chimp brains were also asymmetrical. Yet, chimps have no complex language like we do, so what does this mean? Jim Rilling may have part of the answer. He did a brain-scanning study of language-processing in humans and chimps. He used two chimpanzees that had been involved in ape language projects, and thus were held to have some limited competence at symbolic communication, and scanned their brains after they had undergone a session where they were required to interpret spoken words and symbols representing objects. Human subjects were also required to do the same tasks. He found that humans and 'language-competent' chimps recruited different brain areas when processing these stimuli. Humans activated left-hemisphere areas within the classically defined Wernicke's area and the chimps did not—they used different parts of the frontal cortex, the cerebellum, and the thalamus. He did not necessarily take that to mean the chimps lacked a Wernicke's area, but rather concluded that they must put it to different uses, like interpreting gestures, calls, and facial expressions. It may well mean, he concluded, that humans have indeed evolved unique language-processing brain regions.

Rilling, Preuss, and colleagues subsequently looked in detail at the structure in the human brain which allows Wernicke's area and Broca's area to talk to each other. It is a thick bundle of white-matter fibres called the arcuate fasciculus. They compared the extent of the arcuate between humans, chimpanzees, and macaques and found that, while the tract extended into Broca's area in the frontal lobe in humans, and to its equivalent area in chimps and macaques, it branched much more extensively in the human frontal lobe and extended much deeper, and in a much more complex way, at its other end, into the human temporal lobe. In fact, these temporal-lobe projections were extremely weak in the chimpanzee and totally absent in the macaque. This unique human specialization was widespread in two areas that are heavily involved in the meaning of words as they occur in sentences in language. In other words, the arcuate joins together a much wider network of language-related areas both in the temporal lobe and the frontal lobe. These are the areas of temporal cortex that have enlarged disproportionately in the human lineage, following the divergence of human and chimp lineages, says Rilling.

There's further 'devil in the detail' to come. Manuel Casanova and Daniel Buxhoeveden, of the Medical College of Georgia, looked at the planum temporale and Wernicke's area, with even greater

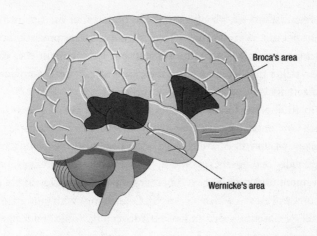

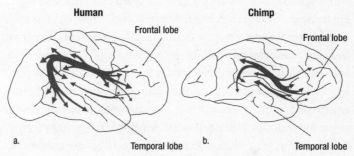

FIGURE 18. The arcuate fasciculus in human and chimpanzee

resolution. They did not look at individual neurons but the pattern into which the neurons are bundled together to form mini-columns. These are bundles of 80 to 100 cells and the connections between them. They found that the human columns were larger, contained more neuropil space (the area outside the actual cell bodies of the neurons which teems with their tangled dendritic projections), and packed more cells into the centre of the column, than those in chimps and other primates. They also found that the columns were larger on the left side of the brain than the right, whereas there were no differences noted for any non-human primate. So there is re-organization—re-wiring—unique to humans, in the side of the brain specialized for language. They are not yet clear precisely how this may lead to cognitive differences, but it is thought that this larger, richer mini-column structure means a great deal more inter-connectivity inside these parts of the brain.

Jim Rilling has found that both the cerebellum and neocortex, which are linked, have enlarged most, relative to body size, in humans. Maybe they have evolved in tandem as a coordinated system? Evidence suggests, says Rilling, that the cerebellum refines both movement and cognition. It 'improves the skilled performance of any cerebral area to which it is linked. Connections with motor areas would increase the speed and skill of movement, while connections with cognitive areas would improve the speed and skill of thought.' For instance, it has been shown to be implicated in memory retrieval, verbal fluency, and the control of attention. If it receives information from the frontal lobes which suggests the body is about to move it will coordinate those movements via its output to the motor cortex. If the cerebellum is damaged, such movements become jerky and erratic. When abnormal it is associated with autism, ADHD, schizophrenia, and dyslexia. You will remember from chapter 2 that the KE family, who suffered from severe language and articulation problems, had an abnormal cerebellum.

The inter-connections between the cerebellum and the cerebral cortex are vast, involving some 40 million nerve fibres—40 times more than the number of optic fibres in the optic nerve. It is a vast computer, massively connected. It is thought that the cerebellum makes predictions about what internal states in other parts of the brain are necessary to optimize their function. As such it can automate action, which is why we can do very complicated things, after a bit of practice, without even thinking about them. Language requires both motor and mental activity. The motor activity is that of speech—the production of vocalizations—the mental part formulates what is to be said, composes sentences out of phonemes, recalls words from memory, and shunts words together to construct meaningful sentences.

For Rilling, it is conceivable that the cerebellum has improved the performance of the prefrontal cortex in ways that have improved apes' and humans' self-awareness, components of theory of mind, and capacity for symbolic thought. But the human cerebellum is not simply a scaled-up ape cerebellum; it is markedly larger even when correcting for body weight and it has been shown that a sub-division of it—the ventrolateral dentate nucleus—is more prominent in humans than the great apes. What tangible advantage could these developments in the cerebellum have handed our ancestors? Rilling identifies accurate overhand throwing, and dexterity in the manu-facture and use of tools. There could also have been clear advan-tages for language, since the cerebellum has connections to Broca's area. I think there is a general principle at work here because accu-rate throwing, deft hand and finger movements, and fast and flu-ent speech and language, all have one thing in common—very fine motor coordination.

When we look deep into the frontal lobes we begin to see a num-ber of structures that *have* changed quite significantly. There has been intense re-organization in the line leading to humans—most of it in the prefrontal cortex, the most anterior part of the brain. The orbitofrontal cortex, part of the prefrontal cortex, is situated over the bony orbits of the eyes. It is critical to judgment, insight, motivation, mood, and emotional reactions—vital for the process by which emotion contributes to appropriate forward-planning and social behaviour. It is the key link between the 'feeling' brain and the 'thinking' brain.

The most celebrated case which demonstrates the importance of the orbitofrontal cortex is that of Phineas Gage. I caught up with Phineas—or what is left of him—in the Medical College Museum of Harvard University, where his damaged skull is on permanent display. He was a construction foreman working on the track-bed

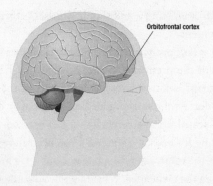

FIGURE 19. The orbitofrontal cortex

for a new railroad in Vermont in 1848. One day he had prepared
a hole in some rock to receive an explosive charge. The gunpow-
der was poured in and he was tamping the charge down when
it exploded. The tamping iron—over three feet long and over an
inch wide—shot through the side of his jaw, traversed the orbit of
his left eye, exited through the top of his skull and landed over a
hundred feet away! Amazingly, he survived the accident, probably
due to the fact that the end of the iron facing him was pointed. However, over the next few months and years, his behaviour degenerated.
He turned from being a jovial, likable leader of men into a fitful,
profane boor, extremely capricious in making ridiculous plans for
his future life that were instantly abandoned. A few years ago, the
neuroscientists Hannah and Antonio Damasio methodically pho-
tographed the injuries to Phineas' skull and re-created the accident
using 3-D computer images. The tamping iron had severely damaged
his left orbitofrontal cortex.

Modern Phineases often result from motorbike accidents. If you
are thrown from the saddle at high speed and come to rest very suddenly when you hit an unmovable barrier, your brain slops around
inside the skull like a wet sponge in a bucket. As it ricochets to and

FIGURE 20. Reconstruction of Phineas Gage's brain injury

fro the orbitofrontal cortex becomes badly lacerated by sharp bony projections on the inside of the skull. I interviewed the survivor of one such accident. He admitted to a range of inappropriate behaviour toward women and disastrous and naive choices in marriage partners. He recalled his second wife turning to him on the steps of the registry office, after the civil marriage ceremony, to say: 'I might as well tell you now that I have no intention of ever sleeping with you—in fact I have no intention of even living with you. I only married you for the green card. Goodbye!' It came as a complete shock to him!

Katerina Semendeferi compared the orbitofrontal cortex across all the great apes, including man. She was particularly drawn to a small posterior area of it called Area 13. This was very large in orang-utans, smaller in gorillas and chimpanzees, and very much smaller in bonobos and again in humans—while the rest of the orbitofrontal cortex in bonobos and humans had a much more complex structure. The intricate structure of the orbitofrontal cortex seemed to mirror a gradient in primates from the solitary, relatively antisocial orang-utan to the complex sociability of bonobos and humans.

The orbitofrontal cortex is crucially concerned with social cognition. It projects backwards to three important, interlinked brain areas: the anterior cingulate cortex, the insula, and the amygdala; and it projects forwards to new and higher-order parts of the prefrontal cortex. This richly inter-connected set of brain organs has been collectively dubbed 'the social brain'. It has been extensively studied in monkeys, apes, and humans and it has recently been discovered that there are a number of important evolved differences at the levels of gross anatomy, internal organization, and cellular structure between humans and the rest of the primates—differences many researchers think reflect important distinctions in social cognition that are unique to humans.

The amygdala is the emotional centre of the brain, reacting particularly to the emotional content of facial expression. Humans with damage to the amygdala are unable to recognize fearful faces and appear not to pay any attention to the eyes, which, wide and staring, give away to most of us the emotional turmoil of fear within.

Recently, Katerina Semendeferi, and her team at UC San Diego, looked in detail at the amygdala across a range of primate species including humans. The human amygdala was much larger than those of the apes, in absolute terms, although as a proportion of the enlarged human brain it was actually smaller. However, the devil, again, was in the detail, because one of the key sub-divisions of the amygdala, the lateral nucleus, occupied a much larger proportion of the amygdala in humans than in any other ape species. It is

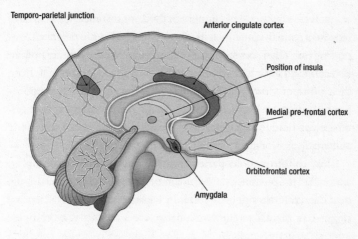

FIGURE 21. The main components of the social brain

this nucleus which has connections into the temporal lobe and the orbitofrontal cortex.

Rebecca Saxe, of the Massachusetts Institute of Technology, believes several aspects of social cognition—'theory of mind'—are unique to humans and reflected by differences in brain structure. For instance, comparative cognitive psychology, she reminds us, tells us that both human infants and apes attend to faces, bodies, and actions, and understand very basic mental states such as goals and perceptions. However, only humans develop a higher-order theory of mind, what is called representational theory of mind, which is the ability to understand the object of a mental state—what a belief or false belief is about—and what the person believes or perceives to be true of it. In other words, to be able to represent the beliefs that are in someone else's head. Saxe believes a region at the boundary between the temporal and parietal lobes, called the temporo-parietal junction, is the seat of our understanding of beliefs.

The temporo-parietal junction reacts strongly to stories involving beliefs and false beliefs, but not to more general description about a character—whether she is hungry or tired, for instance. It has to have representational content. It is also concerned with perspective—putting oneself in someone else's shoes—and in determining how an object would look to a second person, as opposed to oneself, from their point of view. In that sense it is interesting that the temporo-parietal junction has been implicated in out-of-body experiences, where you have the sensation that you have moved out of your body and are dispassionately looking down upon it.

Close to the temporo-parietal junction, in the parietal lobe, is an area called the precuneus. Obscure until recent fMRI brain-scanning, it is currently shooting to stardom because some neuroscientists claim it is pivotal to human consciousness. It has very complex internal structure and a huge metabolic rate—consuming one-third more glucose than any other area of the cerebral cortex. It is also considerably larger, relatively, in humans compared to chimps, and strongly connected to the medial prefrontal cortex, one of the most recently evolved areas in the brain, situated just behind the temple. It is extremely active when you are doing yoga, or when you are in a reflective state, which is why it is considered essential for human self-awareness.

Put this book down for a moment and just sit quietly in your chair—trying to empty your head of all current thought of pre-frontal cortices! After a few minutes your mind will begin to wander. You might call it day-dreaming, but, whatever it is, before too long you will find that you are re-visiting something in your recent past—what happened with the kids at breakfast—why your business partner acted strangely the other day. Visual images will, of course, accompany this mental free-wheeling, and you will 'hear' speech, or narrate past and future events to yourself. Jim Rilling, Todd Preuss, and colleagues have recently reported an extraordinary experiment that compared brain activity between humans and chimpanzees, not when either species was focused on solving some problem, but when their brains were in this 'resting state' of mind-wandering. They found high levels of activity at rest in humans in medial prefrontal cortex (MPFC) and medial parietal cortex. Both areas are implicated in reflecting on the mental states of self and others, recalling memories of self in relation to past events and places, thinking about the future and remembering the past (mental time-travel), and conceiving the viewpoints of others. They also saw high levels of activity in lateral cortex, including left posterior temporal and inferior parietal areas. These areas are involved in our memory of understandings about the world. They also monitored activity in the connection between Broca's and Wernicke's areas: after all, we mind-wander using language—inner speech.

The chimps were similar to the humans in that they showed high activity in MPFC and medial parietal cortices and other parts of the prefrontal cortex. However, there were some species differences. Within MPFC, humans showed the highest level of activity in more dorsal areas, whereas chimps showed more activity in ventral areas. Dorsal MPFC is involved with thinking about others' thoughts as well as one's own, whereas ventral MPFC is more involved with emotional processing. It is possible, Rilling says, that the chimp resting state runs with higher emotional content than the human resting state, perhaps including greater representation of emotional states as opposed to thoughts. Importantly, the MPFC is another brain region that has grown disproportionately in humans.

The thing that stood out above all others was the lack of activity in chimps in all areas associated with language production and processing in humans—the left-lateralized frontal, parietal, and temporal regions. These are involved in conceptual processing and manipulating knowledge for organization, problem-solving, and planning—key advanced aspects of resting-state cognition that clearly differentiate humans from chimpanzees. So, it seems that apes do mentalize—but in a very different way. Their reveries are emotionally charged snapshots, ours are mental movies crammed with complex plots, rich dialogue, future plans, past memories, and a cast of finely drawn characters.

In 1999 came the first report of a new kind of nerve cell, unique to humans and the great apes, called spindle cells. They were discovered by a group of scientists from Caltech led by John Allman. They were first found in a brain region called the anterior cingulate cortex, so called because it forms a collar of tissue around the corpus callosum, the large bundle of nerve fibres that links the two hemispheres. The anterior cingulate is an extraordinary junction that integrates basic information to do with things like blood pressure, heart rate, and gut sensations (the so-called autonomic nervous system), and emotional stimuli like facial expressions and the sensation of pain, with much more sophisticated higher-order cognitive functions such as attention, decision-making—especially in the face of conflicting information—anticipation of rewards, vocalization and language, and empathy. In fact, the anterior cingulate is activated during the experience of any intense emotion, be it anger, love, or lust. Intense sensations such as pain, hunger, thirst, or breathlessness all trigger it. As such it is strongly connected to the amygdala, another part of the 'social brain' called the insula, and many regions of frontal and parietal cortex.

Allman and his colleagues compared the populations of spindle cells in the anterior cingulate across the great apes and found that they mirrored the social complexity of higher primate society. They were very rare in orang-utans—representing fewer than 0.6% of all pyramidal cells; slightly more common in gorillas at 2.3%; more common still in chimps at 3.8%; and commonest of all in bonobos and humans where they represented 4.8% and 5.6% respectively of all pyramidal cells. In humans they appear in clusters, typically of three to six cells, whereas in the chimp they are found either singly or, at the most, in twos. By 2006, the group had expanded the species list of animals equipped with spindle cells to include a variety of cetaceans:

humpback whales, fin whales, killer whales, and sperm whales. This is clearly a case of convergent evolution because the common ancestor of apes and cetaceans lived over 95 million years ago. Spindle cells may be particularly important for processing and spreading complex social/emotional reactions throughout the relatively dispersed different parts of the cortex in larger-brained animals. They are the long-range transmitters of the brain. As you go from orangutans to humans, not only does the population of spindle cells increase, but their individual cell bodies get larger, and the larger the spindle-cell body the denser and bushier is the network of axons or branching connections between cells, leading to the suspicion that they have widespread long-range connections to other parts of the brain.

Allman is most interested in one of those connections—by which the anterior cingulate cortex outputs information, via the spindle cells, directly to the most anterior part of the frontal lobe of the brain: the medial prefrontal cortex. This link, for Allman, goes to the heart of social intelligence. MPFC pulls out memories from past experience and uses them to plot 'next moves'. In particular, it becomes active when you present people with moral dilemmas in which their decision will directly affect the lives of others. The link from the anterior cingulate, he says, conveys the motivation to act, particularly the recognition of having committed past errors, and allows you to learn from past mistakes. This capacity, says Allman, is related to the development of self-control as an individual matures and gains social insight and experience. In fact, the anterior cingulate shows a steady increase in metabolic activity as an individual progresses from childhood into adulthood. Since the most complex powers of social intelligence are processed here it comes as no surprise to find that MPFC is larger, absolutely and relatively, in humans than in the great apes, and, in fact, declines in size in the same order, from humans through to orang-utans, as the population of spindle cells in the anterior cingulate.

In 2003, John Allman expanded spindle-cell country in the brain to include the insula. This area is adjacent to the orbitofrontal cortex, where, as we have seen, complex social cognition such as forward planning and appropriate social behaviour is processed. They counted a massive 82,855 spindle cells in human insula 16,710 for gorillas; 2,159 for bonobos; and a paltry 1,808 in the chimpanzee.

Very little was known about the insula, until recently, because it is hidden away in the centre of the brain, covered by the lids of the temporal and parietal lobes, and difficult to reach. Only with the arrival of fMRI brain-scanning technology has it been rescued from obscurity. It has rich connections with the amygdala and other parts of the limbic system, and onwards to the anterior cingulate cortex and thence to the prefrontal cortex. Like the anterior cingulate cortex, it is now seen as a crucial processor of very basic information from a vast variety of receptors on the skin, throughout the gut, and in the respiratory and cardio vascular systems, which it converts into subjective feelings. It is where a bad taste in the mouth, or the sight of something execrable, is turned into the feeling of disgust, for instance. The anterior insula, together with the anterior cingulate, is also involved in feelings of empathy toward others because both are particularly responsive to cries of pain and the sight of other individuals in pain or anguish. The insula also becomes active when you crave another cigarette or drink, or when a junkie is desperate for his next fix. It registers being given the cold shoulder in a social situation, leading to feelings of rejection and embarrassment. For the neuroscientist Antonio Damasio, it is the place where mind and body meet; where information from the viscera becomes visceral reactions—literally gut feelings. These sensations are represented in the anterior right insula, which has expanded hugely in apes, and yet again in humans.

All these differences in anterior cingulate, insula, and prefrontal cortex, unique to humans, take on real importance because John Allman believes they are the neurological substrate for moral intuition. He proposes that moral intuitions are part of a wider set of social intuitions that guide us through complex, highly uncertain, and rapidly changing social situations and become more fine-tuned as we gain experience through life. You only have to remember the dramatic changes to Phineas Gage's social competence to realize how much damage can be done to moral intuition and social judgment when one of these important components of the 'social brain' is badly compromised.

Allman proposes that the prime input for moral intuition is the insula. In all mammals, he says, it contains a neural representation of the motor and sensory systems involved in the ingestion and digestion of food. It is thus responsible for the regulation of food intake by registering hunger, thirst, or a craving for a particular foodstuff, and the rejection of toxins through registering bad or bitter taste or smell and turning that into an aversive feeling of disgust, possibly accompanied by the need to retch and vomit. (In primates, and especially humans, he says, there is an additional set of inputs arising from the body that signal sharp pain, dull pain, coolness, warmth, itching, and sensual touch—body awareness across many fronts.) This regulation of food intake is expressed in polar opposites of lust and disgust—literally bad taste. This lust–disgust polarity, he believes, acts as an evolutionary template for a range of complex social emotions which also express as polar opposites: for instance, love and hate, trust and distrust, truthfulness and deception, and empathy versus contempt. The first of these pairs generally promotes social bonds, the latter disrupts them. So the two poles are prosocial and antisocial.

You can see both visceral reaction and moral intuition develop in parallel in young children, says Allman. Children have simple tastes—sweet, salty—whereas adults have more complex tastes—bitter, tannic. Moral intuitions also start out as simple, polar, black-or-white 'right or wrong' feelings and develop shades of grey as we get older. Not only is this reflected in the metabolic activity of the social brain, but the appearance of the spindle cell population in both insula and anterior cingulate has a developmental aspect to it. In humans, the spindle cells do not appear until week 35 of gestation and at birth only 15% of the mature number are present. This sets up the speculation that spindle cell population might be severely truncated either by medical insults during late foetal development or by a badly deprived early childhood—leading to sociopathy later on. Also, spindle cells are more numerous in the right hemisphere, which is specialized for social emotion, and are selectively destroyed in fronto-temporal dementia, which is characterized by difficulties with moral intuition, self-awareness, and bizarre humour.

This whole neurobiological system is all evolutionarily modern, asserts Allman. First there are the inputs to the insula in primates, enlarged in humans, dealing with sensations of pain, coolness, warmth, and touch. Second is the emergence of a novel circuit component—the spindle cells and their great elaboration in humans. Third is the disproportionate expansion in humans of the medial prefrontal cortex—specifically implicated in moral decision-making in emotionally charged situations. To this you can add the human-specific specializations in the amygdala. Altogether an extraordinary engine in the brain, turning visceral sensations, and the sight of pain and facial expressions, into emotional feelings and finally complex social planning and appropriate action, from 'biting your lip' to prevent a conversation taking an embarrassing turn, to planning your love-life with a view to marriage, or deciding what course of action to take, mindful that it might cause harm to another. Gut feelings are transformed into moral intuition.

When we interact with other people we are able to effortlessly and instantaneously grasp why they are doing a certain thing, and what they feel about something or other, because we can, in effect, read their minds. We immediately recognize and interpret complex emotions on their face, and we automatically explain their actions in terms of beliefs, desires, and intentions. We seem to have an intuitive grasp of what others do and feel—we don't really have to think about it as we would if we were busy trying to solve some knotty mathematical problem. How often, in conversation with another person, are you suddenly aware that you have crossed your legs, as they have just done, or raised your arms and linked your hands behind your neck, or scratched your forehead, in a mirror image of that same action in your companion? Psychologists call it the chameleon effect. People who make the best chameleons in terms of such bodily actions turn out to be the most sensitive and empathic toward others. In all these social phenomena we understand the actions and emotions of other people because we can experience the same actions and emotions in ourselves. Something seems to be linking first and third person. We can immediately pantomime someone else's actions, whether comically to lampoon them, or in all seriousness, because they are showing us how to delicately carve a piece of wood into a sculpture with a chisel and we need to accurately imitate their actions in order to do it for ourselves. A growing number of neuroscientists around the world think they have found the neural substrate in the brain for all this linked implicit understanding between people. They have identified populations of neurons in parts of the frontal, temporal, and parietal lobes which they call 'mirror neurons' because they are activated when we see a certain action being taken by another and when we do the same action for ourselves, or when we witness a certain emotion in another, and feel it in ourselves. Observation and execution appear to be served by dual function in the same nerve cells. We are capable of all this social cognition, not because of complex higher-order interaction between our visual cortex and myriad different conceptualizing parts of our brain, putting the whole picture together, but by the direct penetration, as they put it, of visual information into the first-person motor knowledge of the observer—directly linking 'he does and he feels with I do and

I feel' in one economical population of neurons. These same neuroscientists believe that the mirror neuron system in humans goes far beyond understanding the actions and feelings of others. They believe it is the foundation of the very accurate fidelity by which we can imitate others—such that masters can pass on new skills to observant students—which explains the exponential rise of culture in our species. They also believe it may explain how human language evolved. It is only fair, at this point, to make clear that there exists a minority of scientists with grave doubts about the very existence of mirror neurons in humans, let alone a superior human mirror neuron system. However, I have elected to go with the majority consensus view.

Mirror neurons were first discovered in the brains of monkeys by Giacomo Rizzolatti, from the University of Parma, and colleagues. They had implanted electrodes into monkeys' brains so that they could pick up the activity of single neurons. They successfully wired up neurons that discharged when the monkey grasped an object or a piece of food. But, to their surprise, these same neurons also fired when the monkey saw another individual—or a human experimenter—grasping for the same thing. In fact, one report claims that one monkey—tethered to its electrode—put the wind up a research assistant who happened to walk into the lab crunching happily on an ice-cream cone. The apparatus sprang into life as the neuron fired! These neurons were recorded in a part of the frontal lobe called Area F5, which is thought to be the exact equivalent of Broca's language area in humans.

Some of the mirror neurons also responded to sounds. They fired at the sight of a peanut being broken, or a piece of paper being ripped. But they also fired at the sound of the crisp crack of a peanut in the dark—so these 'audiovisual' mirror neurons can represent actions independently of whether they have been performed, heard, or seen. In yet another study it was found that mirror neurons fire when a monkey either observes or enacts movements of the mouth, whether associated with eating or communicating. Specifically, about 80% of this class of mirror neurons are 'ingestive' neurons in that they respond to actions associated with taking in food—such as grabbing it in the mouth, breaking it, or sucking it. The remaining 20% only respond to mouth actions needed for communicating—including lip-smacking.

The mirror neuron system of monkeys allows them to understand the actions of other individuals around them because, when they observe an action, the neurons in their premotor cortex—F5—are activated and produce a motor representation of that observed action. This motor representation corresponds to the signal generated in the same neurons when the individual produces the same acts—acts that are already 'in his repertoire'. In this way, says Rizzolatti, visual information is transformed into knowledge. Mirror neurons allow monkeys to run simulations of actions in the outside world inside their own brains. Seeing becomes understanding.

This 'virtual reality' simulation of the outside world inside a monkey's brain seems synonymous with imitation. Yet we know from countless years of observation that monkeys are very poor imitators. Their mirror system seems to be limited to understanding the actions of others without necessarily being able to accurately imitate them. So, if humans also have a mirror neuron system, what are the differences between the mirror system of monkeys and man that can allow us much more powerful social cognition?

With the exception of the occasional obliging patient who has already had his skull opened for electrode implantation to cure things like Parkinson's disease, it is impossible to do single-cell recordings on humans. However, through a series of ingenious experiments using fMRI and transcranial magnetic stimulation brain-scanning, these scientists have satisfied themselves that they can identify active mirror neuron systems in humans. They claim to have found mirror neurons in that part of the frontal lobe that sits just in front of the motor cortex and contains Broca's language area, which directly corresponds to area F5 in monkeys. They have also found them in inferior parietal cortex and in parts of the temporal lobe. A direct match. But how can the human mirror system operate at a higher cognitive level than that in monkeys? Part of the answer may lie in its size. All these three brain areas lie on the borders of the great Sylvian fissure—a deep cleft that you can see running upward from front to back along the side of the brain. They are therefore called perisylvian structures and, when you compare the human perisylvian with its chimpanzee counterpart, it turns out to be the area that has most increased in relative size between the two species. The human perisylvian has ballooned. Another part of the answer has to lie in the other parts of the brain to which the human mirror system is connected—especially the prefrontal cortex, where all the higher-order cognition is done. Apart from anything else this is likely to give the human system a much richer and deeper understanding of the goals of an action rather than simply an implicit understanding of it.

In that regard, the mirror neuron scientists have argued that the human mirror system operates in a much more abstract way—with more degrees of freedom than that of its monkey counterpart. For instance, the human mirror system appears to have a greater under-standing of the context of an observed action and to discriminate between a hand clasping a mug to drink its contents versus to clean it up, firing not only when a hand reaches out with purpose—to grasp something—but when it waves, apparently purposelessly, in the air, so that it responds to miming actions. As one commentator has remarked, 'they respond when someone kicks a ball, sees a ball being kicked, hears a ball being kicked or hears the word "kick"'. Finally, the monkey mirror system can only represent the entire action—hand grasps mug—in one piece, whereas the human system can break down the action into minute time-bites—chop up the action in a much more detailed way so that it much more accurately reflects the minute detail of the observed action. Such fidelity is vital for accurate imitation.

What about language? Is mirror system simulation at the bottom of this also? When you look deep inside Broca's area you find it is very complex. We think of it as being exclusively devoted to the comprehension and the physical production of speech and language, but it is not. It has an incredibly confusing internal organization where phonology (the sound of speech), semantics (the sense of language), syntax (the rules of language that arrange words to make sentences), hand movements, and mouth actions are all represented, suggesting an evolutionary relationship between hand and mouth actions and language. Dave Perrett, of St. Andrews University, and his collaborator Christian Keysers suggest that audiovisual neurons in our temporal lobes 'see' others making the mouth actions of speech—or the babbling in babies that precedes speech—at the same time that they 'hear' our own baby-talk. If these then mapped onto the appropriate mirror neurons in Broca's area we might be able to train our mouths to make the sounds we see others making. This is, in effect, a motor theory of language evolution. Others, like Rizzolatti himself, together with Michael Arbib, of the University of Southern California, put more emphasis on gesture being the important stepping-stone to language. For them, the human mirror system supports rapid imitation of very complex sequences of gestures—pantomime—and pantomime leads to proto-speech, which builds into proto-language in an ascending spiral. Mirror neurons would satisfy one hugely important requirement of language—which is, after all, 'just' a form of two-way communication.

As both mouth movements and gestures are represented in Broca's area there is credible evidence to support the role of mirror neurons in language production. But what about the meaning of language—the semantics? At the level of gestural communication the semantics are embodied in the gestures used, but in spoken language the meaning of words and the mouth, lip, and throat actions necessary to pronounce them are unrelated. How do we make the leap to abstract meaning? It would help if we could show that hand or arm gestures and speech gestures shared a common neural substrate—and there is a little evidence that this is so. Experiments show that the activity in hand motor-cortex increases in the left hemisphere when we both read and talk. In another experiment, volunteers were asked to grab either a small or large object, and simultaneously open their mouths. Lip aperture increased the larger the object grasped. When asked to pronounce a syllable instead of merely opening their mouths, voice level increased the larger the object grasped. Both lip aperture and power of syllable pronunciation increased when individuals observed another person grasping the same objects.

Alternatively, it could be that the understanding of words has developed from the activation of a mirror system related to monitoring the mouth movements we make when we eat. In one experiment, volunteers were wired up so that electrical activity could be recorded from their tongue muscles while they were presented with a range of words, pseudo-words, or bi-tonal sounds containing either the fricative consonant 'ff' which only requires slight tongue mobilization, or the fricative 'rr' which requires an intensive rolling tongue movement. Listening to the 'rr' words produced a much more significant response from the tongue muscles than listening to 'ff' words, and the tongue responded more strongly to words than pseudo-words. In another experiment, signals were recorded from a lip and a hand muscle while the volunteers listened to either continuous prose or non-verbal sounds, or viewed speech-related lip movements, or viewed eye and eyebrow movements. Listening to speech recorded the highest signal but only in the lip and only when the left hemisphere of the brain was stimulated. They reckon this proves that when an individual listens to real words there is an activation of the motor centres in the brain that make us produce the spoken word.

Can self-recognition, which many cognitive scientists believe is an essential prelude to understanding the minds of others, also be a property of mirror neurons? The neuroscientist Vilnayur Ramachandran thinks so. Together with Eric Altschuller and others, Ramachandran has shown, indirectly via EEG recordings taken from the surface of the skull, that autistic children lack a mirror neuron system—which would explain their lack of empathy, theory of mind, language skills, and imitation. This result has been confirmed using fMRI scanning. Autistic children showed no mirror neuron activity in Broca's area—where most of us learn the intentions of others by imitating them.

Mirror neurons, Ramachandran asserts, are also abundant in the inferior parietal lobule—a structure that underwent an accelerated expansion in the great apes and, later, in humans. As the brain evolved further, he says, the lobule split into two gyri—the supra-marginal gyrus that allows you to reflect on your own anticipated actions and the angular gyrus that allows you to reflect on your body and perhaps on other more social and linguistic aspects of your self. Ramachandran's group has found that the angular gyrus is more developed in humans than in other primates. He calls it the 'metaphor centre' of the human brain. Patients with damage to the angular gyrus, otherwise intelligent and mentally lucid, could not understand metaphor and interpreted everything literally. 'How does all this lead to self-awareness?', asks Ramachandran. 'I suggest that self-awareness is simply using mirror neurons for "looking at myself *as if* someone else is looking at me". The mirror neuron mechanism that originally evolved to help you adopt another's point of view was turned inward to look at your own self. Introspection.'

Christian Keysers claims that complex emotions like guilt, shame, pride, embarrassment, disgust, and lust are based on a uniquely human mirror neuron system found in the insula. So the sight of a social brush-off will give rise to the pain or humiliation of rejection. Yet John Allman and his colleagues claim that the role of the insula in human social cognition reflects its population of spindle cells.

There is, at present, no explanation of whether or not mirror neurons and spindle cells in the insula can be reconciled. The two research camps do not talk to each other because the spindle cell research is based on anatomy, whereas the mirror neuron research is based on physiological measurement. No one has yet 'held up' a mirror neuron for us to see or demonstrated the firing of a spindle cell in real time. I would not be surprised to find that spindle cells are a special type of mirror neuron, and the general feeling is that, at the very least, the two systems functionally overlap even if spindle cells and mirror neurons are not one and the same thing.

Rebecca Saxe, together with Harvard professor of cognitive psychology Marc Hauser and others, have been looking at the role emotion plays in how we make moral judgments. You will remember Phineas Gage, and the extent to which his social judgment deserted him following his tragic accident with the tamping iron. Antonio Damasio has a number of patients at the University of Iowa with similar orbitofrontal cortex damage and Hauser and Saxe subjected them to a classic test of how people process moral dilemmas, called the Trolley Problem. Here you have to imagine that you are standing beside a set of points on a tram-rail. On the main line, five workers are repairing the track, oblivious to an approaching tram. On the spur, only one man is working. If you quickly switch the points the tram-car will be diverted onto the spur. One man will die, but five will have been saved. Nearly everyone takes the utilitarian route and throws the points—so did the orbitofrontal patients. But now imagine you are on a bridge above the tracks. Five workers are below you, oblivious to the approaching tram; there are no points, but on the bridge beside you is a very large, fat man. If you push him off the bridge into the path of the tram he will be killed but will slow the tram sufficiently to save the other five. What do you do? Well, most of us wring our hands in despair and do nothing—the emotional content of actively taking a life overwhelms cold reason. However, the orbitofrontal patients threw the fat man off the bridge—they stuck grimly to the unemotional logic of utilitarianism!

Marc Hauser and his collaborator John Mikhail have posted hundreds of these 'trolley problems' on the Internet and maintain that the vast majority of all people world-wide—all ethnic groups, male or female, young or old, black or white, Catholic or Protestant— answer in the same fashion. They believe moral judgment is a human cognitive universal. In his book *Moral Minds*, Hauser argues that there is a Universal Moral Grammar hard-wired into human brains in the same way that we believe Noam Chomsky's idea of Universal Grammar is hard-wired—allowing very young children, effortlessly, to acquire language. If he is right, then brain regions such as the amygdala, insula, orbitofrontal cortex, and anterior cingulate cortex, by which we process visceral reactions into emotional feelings, and temporo-parietal junction, precuneus, and prefrontal cortex, by which we process our understandings of the beliefs and intentions of others (mind-reading), will be parts of this hard-wiring.

Many researchers have shown that the sensations registered in the insula, like pain and disgust, bear heavily on our moral judgment. We come to moral judgment intuitively—as John Allman has suggested. Most people, having swiftly judged someone else's actions, when asked why they came to that decision, will find it incredibly hard to justify themselves. They just know. The suggestion is that in normal minds emotion and reason are both involved in making moral judgments—both battle it out in the brain when we try to solve these moral dilemmas. And the regions that become particularly active in this 'on the one hand on the other' dilemma-solving are the anterior cingulate cortex and the insula. In fact, recent research has shown that the insula is heavily active when we try to resolve moral dilemmas by plotting fair or equitable outcomes for all parties. All these areas have undergone substantial changes in either size or internal structure between humans and chimps.

Ah! I see we are back outside the Old Bailey—where we started! This is where I asked you to pretend to be the scales of justice—an act you found easy because you know what they stand for, and you knew how to stand for them. Inside, scores of jurors are coming to decisions guided by their own powers of moral judgment in which the fate of the person in the dock depends not only on logical reasoning applied to the evidence presented to them by counsel and witnesses, but also on a complex intuitive calculus in which mind-reading and emotion play a huge part. Juries can be easily swayed by feelings of disgust at the details of a crime, or descriptions of the agony that must have been suffered by the victim, or the pain of rejection and social humiliation suffered by a spouse in a divorce proceeding. They can also be easily cowed by the authority figures of judge, counsel, or expert witness. Nevertheless, they are an essential part of an extraordinary feature of human culture—The Law—and the principles of justice. The way their brains are working when they make these decisions involves a number of regions that have clearly undergone disproportionate evolutionary change unique to humans. We are back to our two brains—the human brain and the 'mini-me' chimp brain. We have had to dig deep to find Todd Preuss's 'slim pickings'—the handful of brain differences found so far that make us human. This is science at the frontier—the picture contains far more cracks than paper—but over the next few years we may see enough knitting together of neuroscience, cognition, and genetics to serve Todd Preuss up a feast. In the next chapter I describe one evolutionary detective story that neatly ties together changes in brain structure, behaviour, brain cells, genes, and neurotransmitters, not only unique to the human line since the split from the common ancestor with chimpanzees—but to the last 40,000 years or less. We humans are the agents of our own evolution—we are the apes that domesticated ourselves.

CHAPTER 11

The Ape That Domesticated Itself

O ver the last few years, several pieces of evidence have come together that have dramatically changed our perception of the rate of evolution in the human genome, when it happened, and what processes brought it about. It is impossible to over-emphasize the importance of these recent developments because they not only show that many of the genetic differences between humans and chimps have happened recently—within the last 40–50,000 years—they show how we humans have brought about new and intense selection pressures upon ourselves that have resulted in these genetic changes. We have been the main instruments of our own evolution.

The first explicit clue to what appears to have been going on came in 2006 from a group of scientists at the University of California at Irvine: Eric Wang, Greg Kodama, Pierre Baldi, and Robert Moyzis. Looking for clues to positive selection in genomes is detailed, forensic work and this quartet see themselves as forensic scientists, searching the human genome for 'Darwin's fingerprints'. In the same way that ultra-modern techniques of forensic pathology can reveal the clues in minute quantities of DNA that lead to successful prosecution,

painstaking application of ultra-modern statistical tools for interpreting variability in human genomes has revealed that Darwin's
'dabs' are all over it and that they were left there very recently.

Here is a riddle: What have humans and maize got in common?
Robert Moyzis and his colleagues have produced the answer. They
first came up with a model that starts with the simple observation that any allele (variant of a gene) that is being selected for will
increase in frequency in the population. If that increase in frequency
is rapid you will find that DNA sequence upstream and downstream
of the selected gene has been carried with it, surviving recombination among chromosomes through successive generations of meiosis. This 'hitch-hiker' effect creates blocks of genes that hang together
because the endless cycles of cut-and-paste by which chromosomes
exchange genetic material have not yet had time to break them up.
This is called linkage disequilibrium. It should start at relatively
high values for the immediate area surrounding a gene under selection, and decay over time. Using a huge database of 1.6 million single nucleotide substitutions (snips) they looked to see whether the
sequence between each snip bore a strong or weak signature of linkage disequilibrium. If the signature was high—meaning that the two
snips and the intervening code have remained intact—it suggested
novelty.

They calculated that approximately 1.6% of the snips exhibited 'the
genetic architecture of selection.' That corresponds to some 1,800
genes and that is 7% of the total number of genes, by the latest estimates, in the entire human genome. They were able to put approximate dates on these events and this means that *at least 7% of the
human genome has evolved within the last 50,000 years!*

The most important categories of genes that had undergone this
recent selection were to do with diet, disease, life-span, and behaviour. The high proportion of protein metabolism genes, they concluded, might have something to do with profound changes associated with a move toward a diet rich in cereals and vegetables. Changes

in DNA metabolism genes, they thought, might be connected to increased life-spans between 50,000 and 2,000 years ago. Modifications to the immune system, increases in tumour suppression, and DNA repair were also 'likely components of our unique human longevity'.

So much for diet, longevity, and disease resistance. What about behaviour and the function of the central nervous system? The serotonin transporter gene SLC6A4 had undergone selection. This gene controls the concentration of serotonin in the synapses between nerve cells. Serotonin is very involved with mood and temperament and we shall be returning to this gene with a vengeance a little further on. Glutamate and glycine receptor genes had also evolved: these are involved in speed of transmission of nerve impulses. They also picked up the controversial brain-building gene, ASPM, thought to be vital for growth and increase in size of the cerebral cortex, and there was also evidence of recent selection for the dopamine receptor gene DRD4, and for G6PD, which is involved in resistance to malaria.

Now, here is part of the answer to the riddle about the similarity of humans to maize:

> Homo sapiens have undoubtedly undergone strong recent selection for many different phenotypes, including, but certainly not limited to the general categories we have listed in this work. Such selected events are not rare. The numbers obtained, however, *are similar to estimated numbers obtained for artificial selection by humans on the maize genome. Given that most of these selective events likely occurred in the last 10–40,000 years, a time of major population expansion out of Africa, followed by regional shifts from hunter-gatherer to agrarian societies, it is tempting to speculate that gene–culture interactions directly or indirectly shaped our genomic architecture.* [emphasis added]

Put simply, Moyzis and his colleagues are arguing that we domesticated ourselves on much the same scale that we domesticated our main cereal crop! Around 60–50,000 years ago, *Homo sapiens* migrated out of Africa and began to populate much of the globe. Around 14,000 years ago, in the Fertile Crescent, the first moves were made from a hunter-gatherer existence to the earliest agricultural settlements. Domestication of cereals and other plants, and cloven-hoofed animals, also occurred at around that time. The advent of agriculture led to changes in diet which dramatically changed the demands made upon our digestive systems; the greater population density inside the new encampments, settlements, and villages led to exposure to new pathogens; and material culture also changed at that time, all imposing a new set of selection pressures that have peppered our genomes with Darwin's dabs today.

Hot on the heels of Robert Moyzis and his colleagues came another report of recent positive selection in the human genome, this time from Benjamin Voight, Sridhar Kudaravalli, Xiaoquan Wen, and Jonathan Pritchard, from the University of Chicago. They had followed the same approach as that of Moyzis' group and compared the three human groups in the HapMap dataset: Yoruba, East Asians, and Europeans. They collected evidence for selective sweeps that are so recent that they must have come about since these three groups parted company. We are talking about genetic events within the last ten thousand years—covering the Holocene period of major human global expansion and agricultural development.

They recorded the recent evolution of genes for detection of tastes and odours, egg and sperm production, bone structure, and fertilization. They also identified genes involved in the metabolism of carbohydrates, lipids, and phosphates, and vitamin transport. Four genes involved in skin pigmentation showed clear evidence of selection in Europeans—*OCA2, MYO5A, DTNBP1, TYRP1*. All four genes are associated with Mendelian disorders that cause lighter pigmentation or albinism. In the evolutionary sense they would have been selected for lighter skin tones in human groups who moved to higher latitudes. These genes were recently selected for—and very strongly.

Now, because I am a blue-eyed boy, these results for OCA2 are particularly interesting—because a group of researchers from the University of Copenhagen have just reported that a mutation in OCA2
between 6,000 and 10,000 years ago gave rise to blue eye colour.
Before that we all had brown eyes. OCA2 codes for P-protein which
is involved in the production of melanin. The mutation did not affect
the coding region of the OCA2 gene itself, but a regulatory 'dimmer
switch' nearby. This dilutes the amount of melanin going into the iris
and 'don't it make your brown eyes blue!'

Voight's group discovered patterns of selection that are peculiar
to each of the three main population groups studied—and therefore extremely recent in human history—as well as genes shared
among all three groups, possibly dating to a time deeper in prehistory. That view was supported by a group of scientists mainly
based at Cornell: Scott Williamson, Melissa Hubisz, Andrew Clark,
Bret Payseur, Carlos Bustamente, and Rasmus Nielson. Using a different statistical method to investigate whole genomes, they analysed
1.2 million human snips in African-American, European-American,
and Chinese samples and identified 101 regions of the human genome
with very strong evidence of a recent selective sweep. In general
they find much more evidence for selective sweeps in the Asian- and
European-derived populations than the African. This fits with the
idea of novel environments encountered by anatomically modern
humans as they radiated out of Africa. They also put a very important
figure to the scale of recent evolution in the human lineage by stating:
'We find that recent adaptation is strikingly pervasive in the human genome,
with as much as 10% of the genome affected by linkage to a selective sweep.'
This exceeds even Moyzis' estimate.

In 2005, a French group of scientists produced a particularly interesting example of the sort of recent evolution we are talking about.
They studied NAT2, which is involved in acetylation, a biochemical process that occurs in the liver to modify and detoxify certain
proteins. They discovered an allele of NAT2 that had been under

positive selection in Western and Central Eurasians in the last 6,500 years. This causes slow acetylation by regulating the activity of the gene. They report that the highest frequency of slow acetylators are observed in the Middle East—where agriculture originated. The emergence of agriculture had suddenly, and dramatically, changed the types of proteins in human diet, presenting the body with a whole range of novel proteins which it interpreted as invading antigens. NAT2 was down-rated to allow the body to accept these new proteins more readily. Interestingly, carriers of the slow acetylation version of this gene today are at greater risk for a variety of cancers, including bladder cancer, and allergic conditions. An adaptation among Holocene agriculturalists has turned into a modern liability.

Watching this flurry of interest in recent human evolution from the sidelines was the anthropologist John Hawks, from the University of Wisconsin. He has always been amused by the view, so prominent in popular accounts of human evolution, that, with the emergence of modern *Homo sapiens* in Europe about 40,000 years ago, and the advent of material culture, we humans had simply stopped evolving in the genetic sense and that any differences noted since then, in stature, physiology, and behaviour, were due to environmental factors—culturally induced. To Hawks, the results obtained by Moyzis, Voight, and Williamson did not come as a surprise. Two well-understood factors to do with recent human prehistory mitigated against evolution grinding to a halt: the explosion in human population numbers since about 40,000 years ago and the explosive change and variation in human habitats and ecology caused by migration and burgeoning human culture. These two factors, put together, he reasoned, should *accelerate* human evolution, not curb it. In late 2007, Hawks teamed up with Eric Wang and Robert Moyzis, and two old pals, the independent biologist Greg Cochran, and the physical anthropologist Henry Harpending, to set out the argument for the recent acceleration of human evolution in clear and compelling detail.

The idea goes right back to Darwin and his thoughts about artificial selection of farm animals. When breeding livestock, he had said, always select from as large a herd as possible. We now know that this is important because the larger the population under selection, the more genomes there are in circulation to bear new mutations in response to any selection pressure. However, in large, static populations, like Darwin's herds, new mutations should become less favoured over time as the population as a whole gets fitter—better adapted to its environment. But we humans are an exceptional species. Human populations, despite a number of nearly catastrophic setbacks, have been forever growing. We have never sat still for long, and have forever found new places to live, and new livelihoods to be had. We have been forever encountering new environments, new foodstuffs, new diseases, and vast changes in our material culture. This is why, by all the recent estimates, between 7% and 10% of our genomes have accumulated change within the last 10,000 years. But to prove that this was caused by the rate of evolution actually speeding up, Hawks and his colleagues had to compare what we see today with what our genomes should look like if the rate of evolution had remained constant over the last 6 million years.

They calculated that if evolution ever since the split from the common ancestor had been occurring at the very high rate we have seen of late, a mammoth total of some 6 million amino-acid-changing genetic substitutions should have built up between us and chimps and, over millions of years, been propelled by selection to fixation—which means all human individuals today would carry the favoured gene variant; an absolutely massive divergence between us and chimps. But our two species are much more closely related than that. Only 40,000 amino-acid substitutions have been documented, of which only a fraction will have been selected for and gone to fixation. This means our evolution must have remained glacially slow for nearly 6 million years and then suddenly, and dramatically, accelerated *one hundred-fold!* The evolution of the lactase gene, they say, is a prime example of their argument. We know that a crucial mutation in this lactose tolerance gene arose between 10,000 and 6,000 years ago. However, when scientists looked for it in Neolithic skeletons dating back a mere 5,000 years they couldn't find it. This was because, although this favourable mutation had arisen, and was being selected for, it remained at very low frequencies for a very long time. Yet today, we find it at over 90% frequency in northern European populations. A dramatic rise.

This extraordinary idea, based on rapid recent rates of evolution, that humans domesticated themselves, is a great theory, but the problem is that we do not really know very much about how crops and animals actually got domesticated—let alone humans. We are not sure when pigs, horses, cattle, goats, and dogs were admitted into the fold, how Palaeolithic humans went about recruiting wild animal species in the first place—if they did—and how long it took for them to tame them. We are even less sure about what changes in these wild species were required. However, the idea that humans did, in some way, domesticate themselves into a highly cohesive, social species that rolled across the globe and colonized all the great land masses of the world is not new. It has been around the block a few times before, and I am indebted to the physical anthropologist Richard Wrangham for suggesting to me that it is high time it was re-explored.

The question for Wrangham is whether or not this principle of self-domestication holds, not just for the last 40,000 years, but as an engine of human evolution spanning key points of transition over the last 6 million. Are we humans some form of domesticated ape? Several important anthropologists, including Franz Boaz and Ashley Montagu, have speculated about it in the past, and the late Stephen J. Gould wrote the final chapter of his seminal book *Ontogeny and Phylogeny* along those lines, publishing a list of what he considered to be relevant neotenic features of humans. These earlier commentators have held that neoteny, or paedomorphosis, has been a major factor in the evolution of domesticates. Ashley Montagu defined it as:

> The retention into adult life of those human traits associated with childhood, with foetuses, and even with the juvenile and foetal traits of our primitive ancestors. It can also refer to the slowing down of the rate of development and the extension of the phases of development from birth to old age.

Neoteny involves the staggering of the developmental clock such that key stages in development from embryo to adulthood occur at different times and last different lengths of time. These changes in developmental timing come about through changes in the timing of bursts of the neurochemicals and hormones that guide this developmental schedule. In terms of behaviour, the ethologist Konrad Lorenz said that the unique and outstanding human trait is that of remaining in an unending state of development. The specialty of humans is non-specialization, versatility. A number of palaeoanthropologists have noted that the juvenile skulls of prehistoric hominids like Australopithecines, *Homo erectus*, and Neanderthals exhibit many features that more closely resemble those of modern man than they do those of their own adult forms. Could modern humans have evolved by the simple retention of the juvenile traits of these forms—neoteny?

Ramon y Cajal, one of the founders of modern neuroscience, noted that the microscopic structure of the newborn human brain more closely resembles that of the embryo than it does the adult. Myelinization—where fatty sheaths enclose nerve fibres—is not achieved until 15 months after birth, and nor are the pyramidal tracts from the brain to the spinal cord which control muscular movement. The immune system takes about a year to kick in and bone development takes at least a year to reach the same degree as in the newborn ape. The human is therefore prematurely born because of the limitations of the pelvis, and should really remain in the womb for another year or so. As a result, infancy and early childhood lasts six years in humans versus three in apes, childhood six versus three years, and adolescence nine years versus less than four years.

This leads to a crucial link between neoteny and human social behaviour. It is this point that Richard Wrangham has picked up on today. The idea was first mooted by the zoologist J. Z. Young, who suggested that:

> With the long, extended period of childhood before puberty, the young would not only be more readily restrainable and teachable by their elders, to the cumulative benefit of the community, but that some of the physical determinants of aggressiveness and non-cooperation might well be eliminated from the population altogether by neoteny. In the evolution of humankind there is little doubt that non-aggressive behaviour has been at the highest selective premium. It is not difficult to understand why, during almost the whole of human evolutionary history, such behaviours have been discouraged. Within the family, the community, and between communities, they would be too socially disruptive.

Young may well have been right, but this is far from hard science. And it has been the lack of scientific support that has caused these intriguing theories about human self-domestication to fizzle out in the past. Did our self-domestication stop at our bodies, immune, and digestive systems? Or, as in the process by which wild animals were converted, over time, into domesticated animals, did it also involve changes in our brains and behaviour? If so, what were they? The big problem is that we have lacked any tangible and persuasive model for what wholesale changes have occurred in any animal thanks to domestication—until recently.

Imagine you are in the countryside outside Novosibirsk, Siberia, and appear to be in the middle of a dilapidated farm, surrounded by rickety wire-netting fences and dishevelled barns. Suddenly an Arctic fox rounds the corner, tail wagging furiously, barking excitedly, and jumps up at you to be fondled and to lick your face, just like a dog. Welcome to an animal experiment so audacious that the handful of Western scientists who know anything about it have dubbed it arguably *the* most important biological experiment of the twentieth century.

This fox—normally so wild it would run a mile from human contact—is a product of an extraordinary breeding experiment that has been kept going since 1957. It was the brain-child of a far-seeing Russian geneticist called Dimitri Belyaev. Belyaev was fascinated by this seemingly intractable question of how a number of wild animal species had become domesticated by man over thousands of years. He decided to see if he could selectively breed tame, docile, domesticated animals out of their wild progenitors, simply by selecting individuals based on their ability to tolerate being approached and handled by a human. He was trying to compress thousands of years of human prehistory into a handful of generations. Belyaev died 23 years ago, after the experiment had been running for 26 years, but his former colleague, Lyudmila Trut, has carried on his work.

As Trut explains, Belyaev was already acquainted with some basic facts about domestication before the experiment began. Under domestication, most animal species changed their body size, leading to either dwarf or enlarged breeds. Their coats, previously evolved to camouflage them in the wild, also changed characteristics. Many domestic species are piebald, lacking pigmentation in some areas; the hair often turns wavy and curly as in some sheep, poodles, domestic donkeys, horses, pigs, and goats; and tails often change too, curling up into a circle, as in many breeds of dogs and pigs. Ears become floppy. This had been noted by Darwin in *On The Origin of Species*. Only one species, the elephant, has droopy ears in the wild. The audacity of Belyaev's experiment was to ever imagine that there could be any causal connection between morphological changes such as these and tameability—the ability to tolerate humans.

Belyaev selected Arctic foxes because he already knew something about them through his earlier fur-breeding days. He acquired 30 male foxes and 100 vixens from a farm in Estonia. His selection criteria, from the outset, were very strict. Over the years, typically no more than 5% of males and 20% of females have been allowed to breed. Fox pups were tested monthly until they reached sexual maturity at approximately 7 months. The experimenter would briefly enter the cage and try to offer each pup some food out of his hand while attempting to stroke and handle them with the other. At 7 months the pups were scored for tameness and divided into three classes. The least domesticated always fled from experimenters and bit when handled. The intermediate ones allowed themselves to be handled but merely tolerated the experience—they showed no friendly response. The most domesticated not only allowed themselves to be petted and fondled, and accepted food, but wagged their tails and whined when the experimenter approached. Only they were bred from.

Things progressed so rapidly that by the sixth generation they had to add a more rigorous class of domestication to accommodate pups that were so eager to establish contact that they would approach the experimenter, whimpering to attract attention, excitedly jumping at him, and sniffing and licking him just like a pet dog. Today, these 'elite' foxes make up 80% of the experimentally selected population. Forty years, and 45,000 foxes after Belyaev's first experiment, the Institute has a unique population of a hundred foxes, the products of around 35 generations of selection.

The researchers noticed a number of strong differences between the domesticated foxes and their wild counterparts. Wild fox pups start responding to sounds from day 16, and do not open their eyes until day 18. The domesticated pups responded to sound two days earlier and started peeping out one day earlier. Wild fox cubs first show the fear response at 6 weeks old—domesticated pups do not show it before 9 weeks or even later. This means that the

FIGURE 22. Dimitri Belyaev with his domesticated foxes

domesticated pups have a head start in interacting with humans and absorbing themselves into the human social environment.

Moreover, says Trut, this delayed development of the fear response is reflected in differences between the two groups of foxes in the level of circulating plasma corticosteroids—hormones concerned with adaptation to stress. In foxes, she says, the levels of corticosteroids rise sharply between the age of 2 and 4 months and reach adult levels by 8 months. But in the domesticated foxes, the longer the fear response was delayed, the later came the corticosteroid surge. Their social behaviour had changed as a result of profound changes in internal physiological and hormonal mechanisms.

Belyaev concluded that the genes he was selecting for exerted their effect on behaviour through changing the developmental sched-ule of the animal, and this perturbation in normal development resulted in the expression of a whole range of seemingly uncon-nected characteristics. He had been tapping into genes that affected crucial neurochemical and neuro-hormonal mechanisms, causing what is known as 'correlated' or 'pleiotropic' genetic effects. More of these were noted as research went on. The domesticated foxes

FIGURE 23. Domesticated fox cubs

showed a steady drop in the activity of the adrenal glands—which release corticosteroids in reaction to stress. Corticosteroid levels halved after 12 generations of selection for tameness, and halved again after 30. The neurochemistry of the foxes' brains had also altered. Levels of serotonin—thought to be an important mediator of aggressive behaviour—were much higher in the domesticated foxes.

The foxes' skulls had also been changing. In the domesticated foxes of both sexes, cranial height and width was smaller, and snouts shorter and wider. The skulls of the males, Trut says, had been 'feminized'. They were no longer slightly larger in volume, and sexual dimorphism over a range of cranial measurements had decreased. Some researchers believe these are neotenic features and that the foxes had followed the same path taken by dogs in their evolution from wolves.

This backwater Russian research institute has not only succeeded in producing foxes so tame they behave just like dogs, they have also bred tame Norwegian rats, otters, and mink. In the process they have not only given us a fascinating and unique insight into how a host of animal species may have evolved to live with man, but they may have told us something incredibly important about our own evolution. Have we selected ourselves for tameness and relative lack of aggression in much the same way these Russian scientists have tamed their foxes?

Richard Wrangham raised this very idea himself a few years ago. He feels that the concept of domestication is now respectable and can be applied to human evolution from the solid scientific base founded by Belyaev. He contacted his colleague Brian Hare, then also at Harvard, mindful of Hare's previous work comparing dogs, wolves, and chimpanzees on their ability to follow human cues as to the location of hidden food. They decided to extend the test to include untamed and domesticated foxes, raised at the Institute in Novosibirsk, in order to try to distinguish between two competing theories to explain the sharpened social intelligence of dogs and their unwavering, empathic attention to humans. Back in the Pleistocene, had humans selected wolves for social intelligence or had these cognitive powers arrived as a correlated trait associated with loss of fear and aggression toward humans?

As you might expect, the domesticated foxes performed extremely well on tasks requiring them to follow point and gaze cues from a human experimenter as to the location of food. They performed as well as dogs and young children, far better than wolves or untamed foxes, and much better than chimpanzees. Their behaviour was instantaneous: no learning over the duration of the experiment was involved. The domesticated foxes were also more interested than the untamed foxes in approaching and playing with toys if they had seen a human gesture toward them. The results strongly supported the idea that dogs had become domesticated in much the same way as had the foxes—naturally selected for their ability to tolerate the proximity and approach of humans. What had emerged was the high level of sociability and social communication with humans we associate with our best friends today.

The Russian research puts paid to a popular theory, espoused by John Allman and Darcy Morey, which suggests humans deliberately selected from wolves by capturing wolf cubs and keeping them as pets. Those animals which were not submissive to humans and retained a large degree of ferocity or ferality, they suggested, would have been killed, thus selecting for more docile animals which retained puppy-like submissive behaviour. That behaviour led to a number of other juvenile traits persisting into adulthood—arriving at the neotenic, domesticated wolf we call a dog.

The biologist and animal behaviour expert, Ray Coppinger, thinks this idea is totally wrong. Coppinger believes that wolves domesticated themselves. It was natural selection not artificial selection. The key change in the environment of wild wolves was human encampments. If a great many wolves over a long period of time, under their own steam, began to find human encampments enticing as a resource, they would begin to do the evolutionary work for you. The resource was access to a regular food supply. It is estimated, says Coppinger, that only two out of twenty-five cubs born to the average wolf female over the course of her active reproductive life will survive to adulthood. That is a 92% rate of attrition. Human camps threw out bones, scraps, hides and skins, rotting fruit, vegetables, grains, and human waste. All of this was invaluable for a breeding wolf. But wolves are notoriously shy when approaching rubbish dumps—to this day. Here the 'fight-or-flight' response really comes into the picture—with its echo back to the neuro-hormone research in Novosibirsk. Coppinger says the crucial variable is 'flight distance'. When interrupted by human presence wolves fly fast and far and do not return for some time. All this takes up valuable energy—quite enough to leave the wolves in negative calorific equity. Here is a crucial basis for selection. Any wolf able to tolerate slightly more of, and run slightly less from, human company would be at an important selective advantage. Immediately selection is operating on neurotransmitters and neuro-hormones and the important hypothalamus and adrenal glands out of which not only the stress response, but cascades of developmental biology, are manufactured and controlled. The wolves were taming themselves and turning into dogs. Eventually this process might have been assisted by some form of dawning realization on the part of humans that the wolf/dogs had their uses. They were good alarm bells, and it might have become obvious that their senses were useful additions to spears and clubs and other weaponry on hunting expeditions.

Although most observers tend to buy Coppinger's Belyaev-informed account of how dogs became domesticated, not everyone agrees with his starting point—the grey wolf. There are competing theories that suggest the gulf between grey wolves and dogs is too wide on a number of counts and that the most likely ancestor was a smaller wolf—Canis lupus familiaris—a bit more like a jackal. However, the key point, that domestication was a process of natural selection, mirroring Belyaev's experiment, is widely agreed upon.

The Max Planck Institute for Evolutionary Anthropology in Leipzig aims to chart the evolution of humans from our common ancestor with chimps in terms of socially intelligent behaviour, and through genes. You will remember the energetic head of their genomics section, Svante Paabo, in chapter 2, fishing for language genes in Oxford and going back to Leipzig with FOXP2 in his briefcase. This time, Richard Wrangham tipped him off that Belyaev's experimental findings could be a short cut to understanding the evolution of human social cognition. In true fashion, Paabo, never one to pass up a good idea, immediately made contact. The idea was to look for the genes that underpinned the neurochemical changes in the brains of domesticates and see if those same genes in humans had evolved compared with other mammals and chimpanzees in particular. Speed was of the essence and so Paabo got Lyudmila Trut to agree to donate a colony of Norwegian rats selected for tameness, and a colony selected in the reverse direction, for aggression. He then handed the project to one of his young researchers, Frank Albert. The difference in behaviour between the two colonies is dramatic. The tame rats will run to the bars of their cage when an experimenter approaches and poke their snouts through the bars to be petted.

The aggressive rats hurl themselves at the bars in balls of screaming fury. In 2007, Frank Albert began a complex series of genetic crosses between the ferocious and tame rats and expects, eventually, to locate what he calls 'the genetic architecture of tameness'. Will the same genes that cause tameness in rats be involved in the domestication of other species—right up to the differences between chimpanzees and humans?

Richard Wrangham has teamed up with Brian Hare to test bonobos and chimpanzees in the Kibale National Forest in Uganda. His hypothesis is that bonobos are semi-domesticated, neotenic chimpanzees, and that ecological differences between the two closely related species can explain the differences in evolution of their social structure and dynamics.

Wrangham began his career working at Gombe under Jane Goodall and the discovery of the chimpanzees' dark side is etched on his memory. In his book *Demonic Males*, written with Dale Peterson, he describes the attacks by the Kasekela colony raiding parties on the neighbouring Kahana colony. The attacks were incredibly vicious, with the males whipping each other into a frenzy of violence typified by screaming, biting, barking, and hooting, as they beat, bit, and stomped their victims to the point of death. At first the researchers found it difficult to square this vicious, murderous rage with the contrasting picture of chimp society which emphasized fun and friendship:

> The new contrary episodes of violence bespoke huge emotions normally hidden, social attitudes that could switch with extraordinary and repulsive ease. The male violence that surrounds and threatens chimpanzee communities is so extreme that to be in the wrong place at the wrong time from the wrong group means death.

Nor is chimpanzee violence restricted to raids on other communities. Within one colony there is continual, sporadic, violence in pursuit of male hierarchical dominance and a great amount of violent coercion of females which involves charging, kicking, hitting, biting, slam-dunking, and rape.

The differences in emotional tone between groups of chimpanzees and groups of bonobos can be dramatic. Chimp parties run on fear and aggression. Tension lies close to the surface. Demonic males make mayhem. You can almost taste the tang of adrenalin in the air; it hangs like petrol vapour at a race-track. Bonobo parties, in direct contrast, are much more good-natured affairs. A lot of it has to do with the importance of female alliances. On the rare occasions, in chimps, when the girls *can* get together, the males tend to be less aggressive. This is always the case for bonobos. Although they do fight, and there are skirmishes between groups, they do not patrol other groups with lethal intent, so aggression within and between groups is much reduced. Bonobos, says Wrangham, are neonatal chimps:

> They weigh, on average, the same as the smallest chimpanzee. Their heads are small. Their bodies are slender. Their arms and legs are long. Their mouths and teeth are small. They have a long, thin shoulder-blade compared with a chimp or a gorilla.

The differences in behaviour between the two species are even more striking:

> The gentleness of bonobo calls is only the first of an extra-ordinary suite of behavioural differences that are now known to divide the two species. Bonobos are a tale of vanquished demonism. As we enter the social world of bonobos we can think of them as chimpanzees with a threefold path to peace. They have reduced the level of violence in relations between the sexes, in relations among males, and in relations between communities.

Wrangham believes the one factor most responsible for all these chimp–bonobo differences is group size. Even when food is short, bonobo groups are seldom smaller than sixteen individuals, often much more, whereas chimps fractiously hover between two and nine. Bonobo parties invariably contain females, whereas in chimp parties females and babes travel separately. Social theory says that group size depends on food supply yet chimps and bonobos live in the same part of Zaire, in the same sort of forest, separated only by the Zaire river. The subtle difference, however, is that bonobos combine a fruit-rich diet (which is similar to the chimp) with a leaf-and-shoot foraging diet similar to the gorilla. So, while they are on the move, they snack off herbs and shoots on the forest floor, which greatly supplements their diet and reduces the cost of moving as a relatively large group. Chimps share their forest with gorillas who efficiently remove the salad, while there are no gorillas in bonobo parts of the forest. This party stability produced female power and clearly mitigated in favour of a neuro-endocrine balance favouring reduced aggression. Bonobos, in Wrangham's words, are less emotionally reactive than chimps—their fight-or-flight reactions are less hair-triggered. The similarity with dogs and domesticated foxes is obvious.

Wrangham and his Harvard colleague, David Pilbeam, have suggested that bonobos could have evolved from a chimp-like ancestor through selection against aggressiveness and this may have paved the way for enhanced social tolerance. Hare, Wrangham, and Michael Tomasello have suggested that a similar process may have occurred in human evolution: that a sustained period of self-domestication over the last 40,000 years has been preceded by domestication processes much earlier on. So, if humans represent chimp–bonobo differences writ large, if humans are self-domesticated chimps, when and why did these pivotal events of domestication happen? Wrangham identifies the advent of *Homo erectus* approximately 2 million years ago as one, and the arrival of *Homo sapiens* as the other. The former event, he feels, must have a lot to do with fire—and cooking. Cooking, argues Wrangham, changes everything. It makes food more digestible by breaking down tough plant cell walls and by breaking down complex carbohydrates and proteins into smaller sugars and peptides. It sterilizes it, breaks down plant toxins, and allows hominids to extract more calories from it. All this might be expected to promote the evolution of smaller teeth, bigger brains, less massive jaw muscles, and smaller guts. At the same time it completely changes hominid foraging activity. Individuals would no longer have to be perpetually on the move, having accumulated food they would have to sit there while a fire is made and the food is cooked. It also changes the dynamics of hominid society in that once you have a mound of cookable food, or delicious food that has just been cooked, you have the problem of larceny at the supermarket or restaurant. Females, says Wrangham, would have benefited from protective bonds with favoured males, to safeguard their culinary labours. Now you have a reason for the beginning of stable male–female relationships.

Humans would have formed tight bands around fire, suggests Wrangham, and the food it was used with. Overt hostility and aggression would have been selected against. The latter would have something to do with the evolution of language. The social communication that began to result could have led to coordinated social action within the group and agreement as to what to do as a society rather than as individuals. Group members who transgressed acceptable means of behaviour by being too greedy, or too violent, could be made to pay the ultimate price. Capital punishment could well have accounted for 20% of all violent deaths, predominantly among males, in simple human societies—a huge selection pressure for more docile temperament. These deaths are not typified by anger—they are the calmer exacting of a price for socially unacceptable behaviour—institutionalized murder.

This selection against aggression would have affected genes involved in the development of these hominids and the timing of the onset of their aggressive behaviours, and this would have favoured the neotenic traits we mentioned earlier. According to John Hawks and his colleagues, that evolution, whenever it began (let us say 2 million years ago) is still going on today—and Richard Wrangham agrees with them:

> I think that we have to start thinking about the idea that humans in the last 30, 40, or 50,000 years have been domesticating ourselves. If we're following the bonobo or dog pattern, we're moving toward a form of ourselves with more and more juvenile behaviour. And the amazing thing once you start thinking in those terms is that you realize that we're still moving fast. I think that current evidence is that we're in the middle of an evolutionary event in which tooth size is falling, jaw size is falling, brain size is falling, and it's quite reasonable to imagine that we're continuing to tame ourselves. The way it's happening is the way it's probably happened since we became permanently settled in villages, 20 or 30,000 years ago, or before.

Not everyone agrees with Wrangham's cooking hypothesis. There is precious little hard evidence for fire, let alone cooking, before 300,000 years ago, though this is not to say the evidence won't turn up. The attraction of Wrangham's hypothesis, however, is that it gears up. You can imagine a scenario involving *Homo erectus*, around 2 million years ago, where the social processes and social selection he outlines could have begun. And, in the same way that human evolution has vastly accelerated over the last 50,000 years, so have these processes of institutionalized control of naked greed and aggression. There is a ratchet effect on genes and culture and both are inextricably linked. Just as there is selection for tameness in the domestication of wild animals, says Wrangham, and just as in bonobos there was a natural selection against aggressiveness, in hominids and *Homo sapiens* there is social selection against excessively aggressive people. Social processes work with genetics to balance a number of types of temperament in human populations. Over-aggressive, violent people are penalized by law, and until recently that could mean death—removal from the gene pool. People who are the opposite, who are prone to depression or stress, are penalized in terms of their social prospects and access to mates. As several commentators have noted, neoteny is equivalent to civilization, and, in support of that idea, the most neotenous humans on the planet are in China, Japan, and other Asian societies where you see social norms of individual subservience to the collective. So here you have modern human society typified by a dynamic tension between violence and aggression on the one hand, and social compliance, curiosity, and the impetus for physical and mental exploration on the other.

Again, this is all intriguing theorizing, but the idea of human self-domestication will need more hard science to enable it to achieve lift-off. To extend the domestication theory from foxes, dogs, and bonobos to human evolution is going to need a sophisticated bio-social theory involving genes, neurochemicals, brain anatomy, responsiveness to environmental change, clear links to human temperament, and clear dissociation from anything seen in chimpanzees. Do we even have the beginnings of a bio-social science of this magnitude? I think we do and I want to conclude this chapter by laying the foundations for it in the evolutionary biology of four extremely important neurochemicals and the variations in the genes associated with them. They are dopamine, monoamine oxidase (MAO), vasopressin, and serotonin. I suspect selection for a range of genotypes featuring variation for this 'big four' has unintentionally occurred in recent human history and prehistory, and I would not be surprised if these neurochemicals feature heavily in Frank Albert's research when he compares his rats to humans.

One of the most variable human genes known is *DRD4*, the gene for the D4 dopamine receptor in the brain. It is particularly active in the prefrontal cortex, the region of the brain involved in higher cognition and attention. In one of the coding regions of the gene lies a sequence of 48 pairs of DNA bases that varies for the number of times it is repeated. Gene variants (alleles) containing between two and eleven repeats of this sequence have been discovered and this variation of DNA sequence results in corresponding variability in the resulting protein, which can contain anything between 32 and 176 amino-acids in this area. The variant that has excited the most interest is the 7-repeat or 7R allele because, while originally associated with the trait of novelty-seeking, it has also been found closely linked with attention deficit disorder—ADHD.

In 2002, a group of scientists led by Yuan-Chun Ding and including Robert Moyzis investigated the evolution of these variants. The ancestral allele of DRD4 is the 4-repeat (4R). All variants between 2R and 6R can be produced by simple recombinations or mutations from this, but the 7R variant is an exception. No one-step change can account for it and the team calculated that to generate the 7R allele from the 4R would require at least one recombination and six mutations—a very improbable event or set of events. The team eventually theorized that the 7R variant may be an ancient gene, like the 4R variant, that remained held in human populations at very low frequencies until it suddenly came under selection and rose dramatically in frequency. They estimate that this surge in frequency of the 7R allele dates to a mere 30–50,000 years ago, and may be even more recent than that.

So, if the DRD4-7R variant came under strong selective pressure around 40,000 years ago, what might that selection pressure have been? The answer may lie in major population expansion and colonization, dramatic new Palaeolithic technologies, and rapid development of agriculture. But how could these three link to a gene variant that imbues individuals with impulsivity, short attention spans, and hyperactivity?

Perhaps, they speculated, individuals with personality traits such as novelty-seeking and hyperactivity were useful in driving this human geographical expansion? Only very recently, in the modern classroom environment, have these traits proved maladaptive and disruptive, leading to diagnoses of ADHD that label young people with a psychiatric condition that needs drug therapy. In the Pleistocene, the restless carriers of the 7R variant could have turbocharged human society, though some form of balancing selection between 4R and 7R would have been needed to prevent human populations being entirely taken over by aggressive, hyperactive frontiersmen!

Henry Harpending and Greg Cochran have carried this speculation into more anthropological detail. Because the effects of the DRD4-7R allele show up more prominently in males, they say, one has to ask what the niche in human societies could be for energetic, impulsive, unpredictable, and non-compliant males. Two hypotheses could explain the world distribution of 7R, they say. One is that it is a phenomenon of human dispersal across the globe. The personality of 7R bearers suggests they would have been in the vanguard of waves of human migration, concentrating 7R in populations far away from their starting point. This might account for the relatively high frequencies noted in South America, and the presence of the allele in south-east Asian and Pacific populations, compared with absence in China, from where these Asiatic populations originally dispersed. Alternatively, 7R males could have enjoyed a reproductive advantage in male-competitive societies, either by grabbing the lion's share of food when they were children, or in face-offs with other males as adults.

During the Late Pleistocene, between approximately 40,000 and 10,000 years ago, modern humans were indeed rapidly colonizing the globe. Many were hunter-gatherers but the transition to agriculture had begun, either through occasional tropical gardening, shifting cultivation, or, where human densities were highest, more intensive agriculture. These different regimes were reflected in different dynamics within society, especially between the sexes. Harpending and Cochran apply Richard Dawkins' idea of 'cad' and 'dad' societies to the human condition. Most hunter-gatherer, foraging societies, they say, are 'dad' societies, typified by both sexes sharing the work of finding food, where men and women live, eat, and sleep together with their children, pair bonds endure, the men are not gaudy and devote little time to making weapons or art and adornment. A prime example are the !Kung Bushmen. In low-density gardeners, however, where most of the work is done by the women, the men, freed from all domestic responsibility, devote themselves to self-decoration, machismo, and posturing, and armed raids on neighbouring groups. These are the 'cad' societies. Examples would be the tribes of Papua New Guinea and the Amazon Basin. Occasionally, however, such tribes would undergo political organization (as Richard Wrangham hypothesized) which mitigated against individual male violence inside the group and also suppressed warfare between groups:

> The archetypes of dad hunter-gatherers are the !Kung Bushmen of southern Africa; the archetypes of labour-intensive farmers are East Asians; and the archetypes of local anarchy are Indians of lowland South America like the Yanomamo. It is probably no accident that two of the best known ethnographies of the twentieth century are titled 'The Harmless People', about the !Kung who have few or no 7R alleles, and 'The Fierce People' about the Yanomamo, with a high frequency of 7R. We suggest that the absence of 7R in East Asia is recent, consequent to the establishment of powerful political institutions that allowed population growth and forced agricultural intensification.

We have no firm idea—as yet—how much, if anything, we owe the 7R genotype for our restless colonization of the New World, our empire building, the march of technology, or the driving forces in our board-rooms and barracks. For the most part we have turned a selective advantage into an illness, as the prescription rate for Ritalin among our school-children can testify. However, a further clue to how possession of the 7R allele might have benefited some of our ancestors has come from work by Dan Eisenberg among the Ariaal tribesmen of Kenya. Individuals with the 7R allele appeared to be better nourished in populations of the Ariaal that are nomadic to this day, but 7R individuals were less well-nourished in Ariaal populations that have settled down and become sedentary. Eisenberg conjectures that, since the 7R allele has been linked with greater food- and drug-craving, it might have been very adaptive in the nomadic setting where a boy or young man with the 7R allele might have been better at defending livestock against raiders, or better at locating good food and water sources. In the van of migrating human populations thousands of years ago these characteristics could have proved invaluable.

Vasopressin is a hormone with a potent role in social behaviour, where it promotes pair-bonding between males and females and increases parental care behaviour—licking and grooming of offspring—among males. It doesn't matter whether you are a prairie vole or a human, when you bond with your partner or cuddle your kids, vasopressin is at work. And it is with prairie voles that the story starts.

Elizabeth Hammock and Larry Young, from the Yerkes National Primate Research Center, investigated the relationship between vasopressin and the wide range of social behaviour in voles, the genus *Microtus*. Prairie voles mate for life; the males stick around when the babies come and show willingness when it comes to looking after the kids. They are model dads. Male montane voles are very different. They do not pair-bond, the males have no interest in bringing up baby, and, except when they physically mate, they are remote and aloof from females. To a very large extent, the degree to which they are social depends on the expression of the vasopressin 1a receptor molecule in the brain. The thing that actually varies, and leads to the behaviour differences, is a simple repeat sequence of DNA called a microsatellite in the regulatory region of the gene that codes for the receptor. The gene goes by the name *avpr1a*. In the highly social prairie and pine voles this sequence is greatly expanded into several repeat blocks, whereas in the antisocial montane and meadow voles, the sequence is much shorter.

Hammock and Young showed that the number of these repeats correlated with the expression of *avpr1a* and with the amount of time males spent licking and grooming their pups. It is also a 'man thing' in that while vasopressin systems turn males on to being interested in females, and being good parents, other systems determine female mating and rearing behaviour. Animals with the 'long' allele of the receptor gene showed a high level of licking and grooming, social recognition, and partner preference, and lower rates of activity in the amygdala, the 'fear' centre of the brain.

Hammock and Young then dramatically shifted their analysis of the role of the vasopressin receptor to human behaviour. The human *avpr1a* gene is also highly variable. One of the gene variants has been shown to have a modest association with autism—which is typified by social disinterest and very poor social intelligence skills. Could vasopressin have played a central role in the evolution of the high level of social and nurturant behaviour we associate with humans? They compared *avpr1a* in humans and chimpanzees and discovered that chimps had a deletion of 360 base-pairs of DNA at exactly the point where the human variant associated with autism differed. They then repeated the comparative exercise with bonobo DNA sequence and discovered that it was much more similar to human.

This is profoundly interesting research. Richard Wrangham, you will remember, has hypothesized that differences in the availability of plant foods between chimpanzee and bonobo societies may have driven the differences in their social structure, making male bonobos less obsessed than chimps with dominance, infanticide, rape, and pillage. Did those ecological differences drive the evolution of hormonal systems affecting social behaviour, through the evolution of genes like *avpr1a*? Frans de Waal conjectures that the loss of DNA sequence in this microsatellite in chimps, compared to bonobos, could mean that the last common ancestor of humans and apes was much more like a bonobo than a chimpanzee. Which would mean that our deepest ancestor was more domesticated than we thought, and that chimps had 'de-domesticated' themselves. However, it seems much more likely to me, given the evidence for differences in social behaviour and neotenic features between chimps and bonobos, that it occurred the other way round and that food and diet drove differences in the brain which then drove the differences in social behaviour and social structures we see today.

Richard Ebstein, of The Hebrew University in Jerusalem, has conducted a study where over 200 students had to play an economic game called 'Dictator'. Players could elect to pursue a ruthless strategy and try to grab all the money, others elected to be more generous, and others were exceptionally generous and gave the whole lot away. When blood samples were analysed for variants of *avpr1a* they discovered that the ruthless money-grabbers tended to have the short-repeat variant of the gene, whereas the more altruistic players tended to the long repeat. Ebstein suggests that the short-repeat version of the gene might affect the distribution of receptors for vasopressin in the brain in such a way as to make individuals feel less rewarded by committing altruistic or selfless acts.

Ebstein, Rachel Bachner-Melman, and colleagues, also looked at the relationship between *avpr1a*, the *SLC6A4* gene (which regulates a neurotransmitter, serotonin), and creative dance. They found that a particular combination of an *avpr1a* variant and a low-activity variant of the serotonin regulator gene, was very common in their test population of creative dance artists. Since serotonin and vasopressin interact in the hypothalamus of the brain to control communicative behaviour, and since serotonin actually increases the release of vasopressin in the brain, they believe they have found a brain-chemistry engine for creative dance in the context of mystical experience, sacred ritual, need for social contact, and openness of expression, rather than the athleticism and muscular coordination that dancers share with other athletes. Serotonin is linked to states of altered consciousness, mystical experience, and 'mind blowing' hallucinogens and the authors couldn't help noticing the link with taking 'Ecstasy', which temporarily floods the brain with serotonin, inducing a heightened sense of euphoria, at night clubs specializing in 'out of your head' rave dancing.

Dance, in the way Bachner-Melman chooses to define it, is that form of the art which contributes to creative, even spiritual experience, and a heightened form of social bonding and communication. It may, she says, have its origins in shamanism, early religious experience, and in the induction of ecstatic trance states, and these two gene variants may have their origins in prehistoric time when mystical dance and shamanistic practices are thought to have become an essential part of proto-religious belief and ritual in early human society. Elizabeth Hammock is going on to look at variability in the *avpr1a* microsatellite in human populations to see if it underpins variability in human personality traits from social gregariousness to shyness. Meanwhile, Hasse Walum, from the Karolinska Institute in Stockholm, has discovered that men carrying either one or two copies of a gene variant of *avpr1a*, called *allele 334*, had a raised likelihood of marital crisis leading to divorce, and lower scores on tests that measure the strength, stability, and affection of relationships with their partners. Men with two copies of this allele were much less likely to have married at all.

Serotonin is a potent neurotransmitter, or signalling molecule, widespread in the brain and central nervous system. It affects mood, aggression, sexuality, and many other aspects of behaviour, particularly the regulation of emotions. Serotonin also regulates early brain development and the ability of adult neuronal networks to change their connections and connection strengths in reaction to environmental change. Because it shapes various brain systems during development, it is very important to the type of temperament, or emotional reactivity, adults will end up with, how they will respond to stress, and how impulsive or aggressive they will be. Belyaev and Trut had found, in their Arctic foxes, for example, that selection for tameness had increased serotonin production in the brains of tame versus wild animals. This link between serotonin, social behaviour, and domestication can now be applied to primates and humans, thanks to the pioneering bio-social research of Klaus-Peter Lesch from the University of Wuerzburg, together with colleagues in Germany and America.

Crucial to understanding this control over a range of social behaviour and social organization is the relationship between serotonin and monoamine oxidase (MAO), an enzyme which breaks down both serotonin and dopamine. When MAO activity becomes abnormal a variety of psychiatric disorders can result, including depression and attention deficit disorder. MAO deficiency is also linked to aggression and violence in humans. There is also hard evidence that the MAO-A gene has, like the serotonin transporter gene, undergone very recent evolution in humans. So both partners in this antagonistic neurochemical system have been strongly selected for.

On arrival at Wuerzburg, Lesch joined a neuroscience department working extensively on the serotonin system of the brain. Serotonin is vital for passing nerve impulses down long chains of neurons. While in the neuron the message passes as an electrical event but it cannot jump the synapse, or junction, between two neurons without being carried across by serotonin as a chemical signal and handed to receptor molecules on the other side, where it is re-converted into an electrical signal to pass along the downstream neuron. It is rather like the role of a car ferry in connecting two sections of coastal road across a river estuary. If serotonin stays in the synapse too long it will cause excessive signalling between neurons and so a serotonin transporter molecule recycles the serotonin by recovering it from the synapse and returning it to the upstream or pre-synaptic neuron for further use. It is our old friend the *SLC6A4* gene that codes for the serotonin transporter. Lesch, and others, discovered that the regulatory region of the gene exists in long and short variations which affect the activity of the gene—how much serotonin transporter it can make. The 'short' allele moderated the amount of serotonin transporter produced, the 'long' allele resulted in more transporter production, which removed more serotonin from brain synapses. This, presumably, is why creative dancers bearing the 'short' allele are prone to 'serotonin euphoria' though the downside of serotonin dysregulation, as Lesch has documented, features anxiety and depression.

This variable region is unique to humans and simian primates; it doesn't exist in New World monkeys. In humans and great apes the short- and long-repeat variations originate from the same specific site on the gene, whereas, in rhesus macaques, the variations come from another site, but still come in two types—short and long. Lesch and his colleagues suggested that the variability of this gene might determine what kind of primate society resulted. They tested their theory out on macaques. There are at least sixteen species of macaque and they show an extraordinary range of social behaviour and social structure, from highly tolerant societies with low levels of dominance and its attendant aggression, to highly aggressive, dominance-hierarchy-ridden, nepotistic societies. The least demonic societies are those of the Tonkean, Barbary, and Tibetan macaques, and the most demonic are the rhesus macaques.

They compared a social grade of seven macaque species, from docile to demonic, for variation in the serotonin transporter gene and variants of the monoamine oxidase-A gene, the key regulator of serotonin signalling, which also has short and long repeats in its regulatory sequence. They discovered that tolerant macaque species only had one version of the serotonin transporter gene regulatory sequence, and only one version of the regulatory sequence for the monoamine oxidase gene, whereas demonic species had several versions. Rhesus monkeys showed the most variation of all in these regulatory areas, varying for short, long, and extra-long repeats for the serotonin transporter and three variants (five, six, and seven repeats) of the regulatory sequence for the MAO-A gene. So, genetic variation in the regulation of both the serotonin transporter gene, and the monoamine oxidase-A gene, do directly relate to the dynamics of primate societies.

The more violent and demonic a society, the more variation there is in these genes. But why? Lesch suggests that this variability is somehow related to 'the complexity of socialization'. It may be that, in a primate society like rhesus macaques, typified by alarming and sudden shifts in dominance, a variety of personalities, each with different stress thresholds, maintains some stability. Lesch, however, notes that rhesus macaques have an extremely wide geographical distribution compared to many other non-human primates. Somehow, he thinks, this 'broader spectrum of aggression-related traits and behaviours' translates into a species inordinately successful at spreading itself geographically, forever extending its range and taking and holding new territory. A colonizing species.

This ought to remind you of humans, and it has been discovered that chimps and bonobos only have one version of the MAO-A gene whereas humans are hyper-variable for both MAO-A and serotonin transporter genes. So, the two primate species—humans and rhesus monkeys—that are similar in terms of their wide geographical ranges are alike in possessing many variants of these two genes. What actual effects does this gene variation have on contemporary human societies?

Humans who lack the MAO-A enzyme are known to be prone to violent, impulsive behaviour and men who carry the short version of MAO-A are also known to be aggressive and violent, especially where alcohol is involved. These gene variants have been around in primates, in several similar forms, for over 25 million years. Bold aggression may well have been much more appropriate and adaptive in earlier epochs of human, and certain simian, histories, though, even then, you could have too much of a bad thing. Nature has been balancing aggressive and more mild-mannered temperaments. This balance has been dubbed the 'warrior versus the worrier' gene.

Over the last 20 years researchers have made great strides in linking these variants in serotonin transporter and MAO-A activity to behaviour, and to specific parts of the brain, particularly the amygdala, anterior cingulate cortex, and prefrontal cortex—which we dealt with in chapter 10—and are collectively dubbed 'the social brain'.

In the 1970s, American psychologist Avshalom Caspi, together with Terrie Moffitt, now at the Institute of Psychiatry in London, began a longitudinal study of 1,000 babies born in Dunedin, New Zealand. They have been monitoring them every two years ever since. In 2002 they reported that they had been interested to find out why some of the male children in the Dunedin study, having been maltreated as small children, had grown up to develop antisocial behaviour which had frequently got them in trouble with the police, whereas others did not. Their investigation implicated MAO-A. Maltreated children who bore the long-repeat variant of the MAO-A promoter gene, which thus expressed high levels of MAO-A, were less likely to develop antisocial problems. Individuals who bore the short repeat had a tendency toward delinquency. They also monitored serotonin transporter genotype and discovered that individuals with one or two copies of the short-repeat variant were more likely to exhibit depressive symptoms, diagnosable depression, and suicidal tendencies in response to the vicissitudes of life, than individuals homozygous for the long-repeat variant.

In 2007, Lesch's group, lead by Andreas Reif, also looked at the relationship between different variants of serotonin transporter and MAO-A, their relative contribution to violence, and whether a previous history of the experience of violence as a child was important. The short-repeat variant of the MAO-A gene was carried by 45% of violent, but only 30% of non-violent individuals. Individuals who had experienced an adverse, violent childhood only went on to become violent adults if the short-repeat allele of the serotonin transporter gene was present.

In 1992, researchers led by Ahmad Hariri had directly linked these serotonin transporter variants to human emotionality. They put a number of psychiatrically normal individuals into fMRI brain scanners and showed them a selection of pictures, including a parade of scary faces. They were looking for the effect these images would have on the brain's 'fear centre'—the amygdala. The amygdala receives images of facial emotion—like fear—and, as we know, projects to the anterior cingulate. Some subjects showed 'normal' activation of the amygdala, while others showed a heightened amygdala firing response. When they looked at the genes they found that those subjects whose amygdalae had blazed had the short variant of the serotonin transporter gene.

In 2005, Andreas Meyer-Lindenburg, of NIMH, re-visited the fMRI scary faces protocol first used by Hariri. Scans showed that the connections between the amygdala and the cingulate were weaker in those individuals who bore the short-repeat allele of the serotonin transporter. This is important because the cingulate acts as a regulator or dampener of amygdala activity. With feedback from the cingulate compromised there was nothing to put the fire out in these more fearful individuals. Those individuals with poor coupling between amygdala and cingulate had a tendency to an anxious temperament and were hyper-vigilant to potentially harmful situations. As one commentator wittily put it, the difference between normal individuals and individuals with poor amygdala–cingulate coupling, on seeing a thick rope lying on the ground, would be 'My goodness. It looks like a snake!' versus 'Oh! My God! There's a snake here—run for your lives!'

Mayer-Lindenburg went further, this time looking at MAO-A. Individuals who showed the highest amygdala activity, linked to the weakest amygdala–cingulate coupling, possessed the short-repeat version of the MAO-A gene. They then brought the prefrontal cortex into the picture, and the circuits by which it connects and feeds back to the amygdala via the cingulate. As we know, people with damage to the prefrontal cortex often have a tendency to make disastrous social decisions. The connections between the prefrontal cortex and the amygdala via the cingulate form a negative feedback loop. So, individuals with the low-activity MAO-A allele cannot control their emotions because the 'thinking' part of the brain cannot override or dampen the 'emotional' part of the brain.

What else do we now know about serotonin and MAO-A genotypes and the social brain? Intriguingly, says Lesch, the brain regions affected by the low-activity alleles of both genes are involved in imitation, from which social cognition and behaviour in the social world has evolved. Some of these regions, he reminds us, contain mirror neurons, which are activated when we do certain actions, or observe them in others, and spindle cells, which are believed to play a role in social bonding. These astonishing conclusions directly link two gene variants to social behaviour and populations of mirror neurons and spindle cells. This is exactly the combination of genetics, psychology, and neuroscience we need to make sense of human cognitive evolution.

Let us take this just one step further. The scientist who knows the most about spindle cells is John Allman. He pointed out that the sheer number of spindle cells in humans greatly outweighs those in bonobos, which in turn outweigh the spindle cell populations in chimpanzees, gorillas, and orang-utans. In other words, they grade according to social cognition or domestication. These cells are particularly numerous in the anterior cingulate cortex, which, as we have seen, is the go-between for the frontal cortex and the seat of the emotions—the amygdala. They are the source, quite literally, of 'gut feelings' and, as one of Allman's post-graduate researchers puts it 'are responsible for making fast decisions in uncertain circumstances in social contexts'—thinking on your feet.

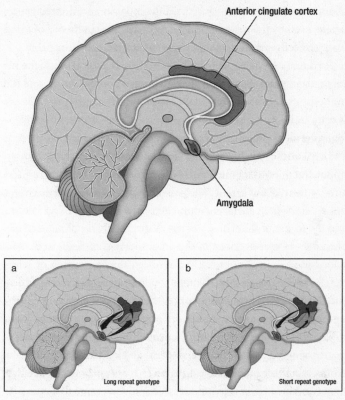

FIGURE 24. Difference in strength of amygdala–anterior cingulate coupling between short-repeat and long-repeat genotypes of the serotonin transporter

Allman has discovered that spindle cells are fairly bristling with receptors. There are receptors for both serotonin and dopamine, and recent research suggests that vasopressin also binds to them! It is tempting to view spindle cells as *the* substrate in the brain where all social intelligence comes together, and, while spindle cells are not unique to humans, they are quantitatively much more numerous. The differential between spindle cell populations in humans and chimps does seem to mirror the differential in social intelligence, where, even by the most optimistic research, human theory of mind vastly exceeds anything in this area of which chimps are capable.

No doubt new research to come will add to the list of neurochemicals and enzymes that are involved in human domestication, and the domestication of a wide range of animals, but I think the groundwork is truly laid for a sophisticated bio-social science that links archaeological evidence of human habitation, stature, commensalism, diet, and agriculture to behaviour—a range of personality types in human populations—and further to neuro-chemistry and the structure and development of our brains. We have not only domesticated ourselves but we have done it more comprehensively than any other species and the magic of the whole process is that much, if not all of it, has been due, as we have seen in the domestication of dogs, to unconscious processes. Domestication has been the by-product of selection for other things. And, though you may argue that all the differences between us and chimpanzees, from variation among neurotransmitter regulators to spindle cell populations and a host of genes to do with the nervous system, metabolism, and immunity, are a matter of degree—quantitative, rather than qualitative differences— I think that these quantitative differences are of such magnitude that their combined effect is to produce a cognitive creature that is unique and whose mind is in a league of its own.

CHAPTER 12

Chimps Aren't Us

Our relationship with the chimpanzee is in need of drastic revision. It has become dysfunctional thanks to a staple diet of popular science articles and television programmes about chimps, and the other great apes, which take a simplistic view of the science available to us and use it to stress their 'extraordinary' similarity to us, in terms of both genetics and cognition, often with the active connivance of senior figures in the world of primatology, who should know better.

There are lumpers and splitters in all walks of life—and primatology is no exception. Scientists like Frans de Waal and Jane Goodall have made careers out of lumping us with chimpanzees. Our understanding of chimps has certainly benefited from their contributions. No-one can forget Goodall's first descriptions of tool use in the wild, and the contrasting accounts of murder and mayhem and care and conciliation that have come out of the longitudinal study of the Gombe chimps. But, as spokespersons for primatology, they have, in my opinion, pandered to and exploited our insatiable anthropomorphism and presided over a 'chimps are us' industry that has gone too far. I much prefer the line taken by another prominent primatologist—Carel van Schaik. While properly recognizing that we can see some very basic foundations for the evolution of many of our advanced human cognitive traits in primate society—as Goodall and de Waal point out—he echoes the sonorous public address system of the London Underground when he cautions 'Mind The Gap!' The cuddlesome 'chimps are us' image is as misguided, wrong, and morally and intellectually bankrupt as the infamous PG Tips Tea advertisements that ran for 40 years on British television—and are now, thankfully, consigned to the dustbin—where a family of jolly chimps, kitted out in jaunty hats and flower print dresses, gathered around a steaming brew of 'Rosy Lee' and jabbered away at each other inanely in Cockney accents. Good riddance to all of you, my Old Chinas, and not lamented!

Yet popular science articles continually report primate cognition research along the lines of 'here is yet another behaviour—once thought to be unique to humans—that we now know we share with the chimpanzee'. One by one, 'unique' human forms of cognition fall, like dominoes. Primatologist Frans de Waal, for instance, takes a positive delight in pointing out that work on morality, tool use, imitation, and altruism in chimpanzees appears to drive nail after nail into the coffin of human cognitive uniqueness. But I believe this gives a false picture of how close the two species really are. When it comes to the difference between humans and chimps I am a splitter, not a lumper. I think the worm has turned. Some prominent psychologists are beginning to talk about human cognition being unique, after all—as a few examples will show.

Take altruism. We know that humans go out of their way to be altruistic. We help old ladies across the road, we give blood, we care and share, and some of us, at least, would, without hesitation, strip off on a river-bank and dive in to save a drowning child. What about chimps? Can we discern the origins of human altruism in chimpanzee behaviour? Some experiments come to pretty dim conclusions. In 2005, for instance, Joan Silk, Jennifer Vonk, and Daniel

Povinelli conducted an experiment in which chimpanzees could pull a tray containing a food reward toward them in such a way that it either only delivered the reward to them, or delivered a reward to them and also to a familiar chimp in an adjoining enclosure. In a second version of the experiment the adjoining chamber was empty. The chimps showed no preference at all in the second version of the experiment—they were just as likely to choose the second option, which also delivered food next door, whether or not there was a chimp present. They showed no 'other regarding' at all—as Silk puts it, it was as if they had simply flipped a coin to decide whether or not—at no cost to themselves (they would get a food reward whatever they did)—they helped the other chimp.

However, two years later—in 2007—a group from Leipzig, headed by Felix Warneken, suggested otherwise: that chimps were as capable as young children of showing spontaneous altruism. Chimpanzees showed strong inclination to help an experimenter, frustrated by an inability to reach a stick, by handing it to her, with or without a reward for doing so, even when the action was at some cost to themselves in that they had to clamber up to a platform to retrieve it. They also showed some inclination to help an unrelated chimpanzee in an adjacent chamber to access a third chamber in which food had been put, by releasing a peg on the wall of their chamber that held the access door between the second and third chambers closed. Although they could help the other chimp, there was no way that they could get at the food once the second chimp had retrieved it. 'The roots of human altruism', Warneken concluded, 'may go deeper than previously thought, reaching as far back as the last common ancestor of humans and chimpanzees.'

However, Richard Wrangham and Brian Hare have recently compared chimps' and bonobos' willingness to collaborate on a task to retrieve food by joint pulling on ropes. Chimps were as likely as bonobos to cooperate with each other to retrieve the food but only if it was in a form they could not easily monopolize—as in a large fruit cut into chunks. If the fruit was whole the chimps would not cooperate. Bonobos, on the other hand, were just as likely to cooperate to retrieve whole or cut food. Wrangham concluded that bonobos are capable of wider food sharing/altruism because their lower emotional reactivity means that they tolerate each other better. If he is right, then bonobos more closely correspond to human altruism, not because of genetic proximity, or shared common ancestor, but because their reduced aggression has resulted from a social structure that is different to that of chimpanzees in that it is constituted of larger groups in which females play a prominent social role and male coalitions are less violent. So, social structure, not genetic relatedness per se, is the crucial factor.

This idea is further borne out by the work of Judith Burkart from the Anthropological Institute in Zurich. Marmosets were given the choice of pulling a tray containing a bowl of crickets, or an empty bowl, into an adjoining area where another marmoset was sometimes present. Only the other individual could get the food—there was no possibility of reward for the 'operator' marmoset. Nevertheless, the marmoset pulled the tray 20% more often when another individual was present than when the adjoining chamber was empty, and did not distinguish between individuals that were 'family' or were not related to them. So here we have a fairly primitive New World monkey sharing altruistic traits with humans. What is common to humans and marmosets, in the absence of genetic proximity, that could explain the similarity? Burkart explains that both species are cooperative breeders, which means that the youngsters are cared for, not just by their parents, but by other adults. In the wild there is a lot of food-sharing between marmosets and theft is tolerated. Human society is typified by a similar assortment of family and non-family 'helpers at the nest'. In these two primates, she says, cooperative breeding was the guiding force for altruism.

Let us have a quick look at imitation. One of the most elegant experiments of the last few years to test chimpanzees' powers of imitation was devised by Andrew Whiten, of St Andrew's University, and several collaborators. It consisted of a slanting perspex box into which a piece of food could be loaded. The food could be retrieved either by lifting a barrier (a bit like raising a sluice gate on a canal) or by poking a flap backwards from a hole in the front, with a piece of tubing. One chimp from each of two groups was taught to lift the barrier, another to poke it. The chimps were then re-united with their group and allowed to continue to reap reward by the chosen method while the rest watched intently. The demonstrations were effective—poking spread throughout one group while lifting spread throughout the other. Chimps presented with the apparatus without the benefit of a successful demonstration performed abysmally. Two months later these two parallel cultural traditions were still being adhered to. These chimps were certainly observing the actions necessary to retrieve the food, and did successfully copy them, showing, at least, how simple food-gathering techniques like ant-dipping could catch on and become long-lived cultural traditions in the wild. But Victoria Horner and Andrew Whiten had gone further into the 'hidden structure' of copying.

They invented another puzzle box for the chimps. In a painted version of the box a flap on the top could be lifted with a stick to reveal a vertical hole which the human demonstrator prodded vigorously. She then turned her attention to the opening of a horizontal tube on one side of the box, and, using her stick, retrieved a morsel of food. It seemed sense, given that the interior of the box was invisible, to assume that poking from the top somehow released the food into the horizontal hole. Indeed, both chimps and young children tended to copy the whole technique—with success. But that apparatus was then replaced with a see-through perspex version. Anybody watching could now see that the vertical hole could not possibly connect with the horizontal tube because of a barrier. So poking away in the vertical tube was totally irrelevant. The chimps figured this out quite quickly, cut to the chase, and copied only the secondary action, which was to poke about in the horizontal tube. In stark contrast, the children persisted in copying the whole routine even after they had witnessed the 'dead give-away' see-through box. Were they just socially conforming—feeling obliged to copy the adult? Horner and Whiten concluded that the children were operating a default rule that said something like: 'Copy the adult faithfully, even if it doesn't seem to make much sense, because it must be important otherwise they wouldn't be doing it'—in other words they were 'mind-reading' that the demonstrator must have intentions behind her actions—and attaching importance to it.

Derek Lyons and colleagues at Yale University carried this idea a little further for children by adapting Whiten's puzzle box into an intriguing suite of puzzles of increasing complexity in which certain actions, pushing or pulling of bolts for instance, were necessary to get at an enclosed toy, but where a range of other actions like tapping on the box or touching a knob were irrelevant. Even on simple versions of the puzzle, the children insisted on copying every action of the demonstrator even when, on separate tests, they had been able to distinguish between necessary and unnecessary actions or had even been instructed to ignore unnecessary actions. They were locked-in to faithful copying of every tiny detail. Lyons calls this phenomenon 'over-imitation'.

So, while chimps emulate, children imitate—even down to the tiniest irrelevant detail. It is not that chimps are actually brighter—immediately cutting through the fiddle-faddle to get to the reward—it is that the two species are learning in a different way. The fidelity with which children seem hard-wired to copy, even at the risk of copying rubbish, becomes very important the more complex the technique that is being demonstrated. With increasing complexity it becomes harder to distinguish irrelevant from relevant actions—the task becomes more opaque—and the only hope is to go for accurate high fidelity. Emulation will begin to run out of steam as it becomes more and more difficult to track the complex actions that eventually lead to a result—whether it be releasing a pellet of food, or constructing a trap to catch game, or honing a flint to make a spear or axe-head. Better to do what kids do automatically—assume that all the adults' actions are causally relevant—even if it makes you look stupid (to a chimp) in the short run or can lead you to copy other people's mistakes. As the authors conclude: 'Indeed, this potent imitation strategy can at times be too potent for the integrity of children's causal knowledge. All of which recommends caution next time you idly fidget with a complex device. You never know who might be watching!'

This difference between emulation and true imitation is crucial to an understanding of how we, as a species, have amassed such variety and complexity of material culture. Understanding that a demonstrator intends his actions to make something, allied to detailed copying of every move he makes, allied to the reciprocal understanding in the demonstrator's mind that he knows something you don't and therefore has to teach you it, produces a potent ratchet effect. New techniques rapidly and accurately catch on and can be disseminated and subsequently modified and replaced by better tools and techniques to do the job in a constant gearing-up of complexity and knowledge that has led from the invention of the wheel, less than six thousand years ago, to the wheeling out of the latest passenger jet.

The general point is that there are clear distinctions between the cognition of primates and humans that have allowed us to build on very modest foundations and blow them up to extraordinary dimensions of power and complexity. Some commentators prefer to categorize these differences as quantitative—more of the same kind—in order to collapse the idea of human uniqueness and narrow the perceived gap between us and our fellow primates. But it is like comparing the abacus with a modern desk-top computer. We see it in the difference between chimpanzee tool use and modern human technology; between primates' sense of social fairness when they punish cheats and share food, and human moral and ethical

behaviour, institutions of justice, and widespread and truly compas-
sionate altruism; between chimpanzee alarm calls, gestures, pant-
hoots, and very simple sentences constructed on laboratory lexigrams, and human language and literature. We see it in the difference between some very basic—though important and interesting—
numeracy in chimpanzees and other primates and the abstractions
of modern mathematics that allow us to guide inter-planetary space
vehicles and probe the basic, fleeting particles of matter. As Michael
Beran, from Georgia State University, has put it: Both human and
animal brains have evolved to deal with numeracy. It is clearly adaptive for a primate to be able to distinguish between a tree bearing ten
fruits, and one bearing only six, or between a group of six predators
on the horizon, versus only three, though that advantage diminishes
in the distinction between, say, 24 and 28, hence the fact that these
are above the limits of primate numeracy. We humans map symbols onto these quantities, abstract them, and manipulate them in
advanced mathematics.

So also for theory of mind. We can allow that chimpanzees and
other great apes can interpret something about the goals of another
animal from its actions, and can understand something about attention and the perspective of another. But humans, above the age of
three or four, understand a lot about the intentions hidden in the
heads of others, and can form propositions about other people's
beliefs and false beliefs. Chimps certainly cannot. There is also an
important link between language and theory of mind—specifically
the ability to represent propositions about belief. A number of scientists, particularly Jill and Peter De Villiers from Brown University,
have argued that language is an essential prerequisite for the representation of beliefs, false beliefs, thoughts, and feelings. These scientists have shown that this higher-order aspect of theory of mind
develops coincidentally with the ability to use complex sentences
involving mental verbs and their complements, as in 'He thought he
saw a unicorn'. Their research with orally taught deaf children, who

have severe language delays, showed that they were also impaired in classical tests of false belief. When their language caught up with their peers, so did their social intelligence. Without language, then, chimpanzees can never bridge a genuine cognitive discontinuity between the two species.

This is a good point to pause a moment on the thorny subject of language. I suspect many readers will feel I have given it short shrift, or, indeed, no shrift at all. What about all those years of ape language experiments? Washoe, Nim, Koko, and Kanzi? No subject has so beguiled the general public while leaving the majority of psychologists and linguists stone cold. For the most part it is a long and sorry history—almost a pathology of science—ridden with wishful thinking, over-exaggeration, and even downright fantasy. Only one study, that of Sue Savage-Rumbaugh and her various colleagues with a number of bonobos, principally the famous Kanzi, emerges with any credit for a determined attempt to probe the limits of an ape's comprehension of symbolic communication. But even here there are criticisms that Kanzi has never actually learned that the lexigrams he uses to make sentences are symbols, though he has learned that certain combinations of them are instrumental in communicating his desires for food and play. True language is specific to humans. I urge readers who want to form their own conclusions on the ape language experiments to look at the devastating critique of the whole field in Joel Wallman's book *Aping Language*, the critical articles of Mark Seidenberg and Laura Pettito, and balance those against Savage-Rumbaugh's own account of her work in *Kanzi: The Ape at the Brink of the Human Mind*.

If language is a true discontinuity between chimps and humans, are there any others? The cognitive psychologist Marc Hauser, after years of fascinating research into the limits of primate cognition, in numeracy, understanding of goals and intentions, understanding of the physical principles of the natural world, and a sense of morals, recently presented a list of traits he thinks make humans unique.

Collectively, they form what he calls his 'humaniqueness hypothesis'. 'The main thesis', he states, 'rests on a paradox. On the one hand, new studies of animal mental life reveal a number of critical building blocks upon which human cognition evolved. On the other hand, the cognitive rift between animals and humans is monumental.' The difference between humans and chimpanzees, dolphins, elephants, and parrots, he claims, is wider than the rift between all these socially intelligent species and earthworms! Humans, says Hauser, uniquely evolved several distinctive cognitive capacities. They are: the ability to combine different sources of information and knowledge to create new insights; to use rules and solutions of one specific problem in a variety of new contexts—i.e. to generalize; the capacity to convert analogue representations into digital symbols, for instance turning quantities in the natural world into numbers; novel brain circuitry for combinatorial and recursive operations—like making tools with several components and materials to do a range of jobs; and an ability to detach modes of thought from raw sensory and perceptual input, allowing us to retrieve from memory in new situations, and to ruminate.

For psychologist Michael Corballis, it is recursion that separates us from all other species. Recursion is a powerful phenomenon that crops up in many branches of cognition. In language, grammatical rules use recursion to generate an infinite number of sentences from the words made up of only 26 letters. Little children learn the power of recursion through stories, he says, quoting 'The House That Jack Built', which builds up through several layers of recursion to 'This is the dog that worried the cat that killed the rat that ate the malt that lay in the house that Jack built'. Theory of mind, says Corballis, is another form of recursion because one is imagining what may be going on in the mind of others. Recursion here can get quite complex as in 'she thought that they thought that John thought that Raymond thought that ...' So also, he says, is episodic memory, otherwise known as time-travel, where you can recall a succession of events involving yourself at different periods of time in the past—or similarly project forward as when estimating the manifold repercussions of some projected plan of action. Recursive rules allow us to count, allow us to program computers with powerful software where routines run sub-routines which run sub-routines and store documents inside folders, inside folders... Finally, we see it in tool-making, where one tool is used to make another tool, and so on, before the resulting implement—an Achulian hand-axe for example—is used on raw material. (I cannot resist reminding you here that, disputed as they may be, episodic memory and recursive tool-making have been claimed for corvids—not chimpanzees!)

All these recursive cognitive phenomena, Corballis points out, are housed in the frontal lobes, which have grown dramatically, in absolute terms, in humans, and where there are strong relative differences in some principal sub-components and in neuron type, number, and organization. Corballis cites the work of Patricia Greenfield on cognitive development in very young children. Their grasp of the hierarchical structure of language and the manipulation of objects occurs at the same time, so that as they begin to combine words into phrases and phrases into sentences they are also combining nuts and bolts into simple structures which are then elaborated in a recursive way. Most children's toys, like Lego and Meccano, work on these principles. Both of these activities, says Greenfield, occur in Broca's area, which we know has evolved in size and internal organization in humans. People with Broca's aphasia can understand what words mean but cannot assemble words into sentences and push them out as speech. They are also very poor at reproducing drawings of hierarchical branching tree structures made of lines. A few quibbles aside, all these characteristics—theory of mind, the concept of a knowing self, episodic memory, mental time travel, making tools that make tools, and counting—are all unique, says Corballis, because of the human capacity for recursive thought.

In late 2007, three more voices were raised in favour of human cognitive uniqueness. Daniel Povinelli and Derek Penn, from the Cognitive Evolution Group of the University of Louisiana, have joined forces with Keith Holyoak from UCLA. Povinelli's group have long argued that humans are alone in being able to interpret the world in terms of unobservable entities like mental states and causal forces, and Holyoak argues that humans are unique in their powers of analogy which lie at the core of scientific heuristics, poetic metaphor, and causal reasoning, and our ability to represent the real world in abstract concepts such as the qualities of sameness and difference, words and numbers, and thus manipulate and exchange complex mental models of the world interpreted as symbols. After an exhaustive review of comparative cognitive psychology to date they conclude that Darwin was wrong to downplay the differences in the cognition of humans and other animals as ones of degree and not of kind, and that comparative research over the last quarter-century has preferred to over-emphasize the continuity between human and animal cognition and exaggerate the idea of a cognitive continuum, rather than explain the discontinuity. They conclude that whatever the 'good trick' was that caused humans to be able to re-interpret the world in a symbolic-relational fashion 'it evolved in one lineage—ours. Nonhuman animals didn't (and still don't) get it.' The important challenge for cognitive science, they say, is to accept this fundamental discontinuity and explain how it could have evolved in a biologically plausible manner.

The most important over-simplification I have set out to challenge is that humans and chimpanzees are extremely closely related genetically—based on the idea that only about 1.6% (or even less) of our respective genomes differs at the level of nucleotide sequence in DNA. The take-home inference from this much trotted-out mantra is that this surprisingly small genetic difference will somehow be reflected in a handful of genes which we will discover to be 'the genes that made us human'. However, generations of genome scientists, from King and Wilson in the 1970s, up to the present day, know this cannot be the case. We now have a huge—and growing—list of structural changes that are known to eclipse the simple picture of single point mutations in the genetic code as the cause of evolutionary change in genomes. These include deletions, inversions, copy number differences, and splice variants. Differences in the timing and rates of expression of identical or similar genes, and the role of 'master-controller' transcription factor genes further act to amplify the genetic distance between humans and chimps. When added into the mix these mechanisms can drive the chimp and human genomes apart dramatically—copy number differences alone accounting for a massive 6.4% of difference, while changes in the immune system genes cause local genome divergence of the order of 13%! And, as we now know, some 10% of the human genome has evolved very recently—with human culture supplying the selection pressure.

In chapters 2 and 3 we discussed a number of examples where human psychiatric pathology had serendipitously provided clues to human cognitive evolution, and it is to psychiatry that I want, briefly, to return. In 1988 I produced and directed a science documentary film for the BBC's *Horizon* series called 'Playing With Madness'. The title was meant to imply that evolution—natural selection—was playing with madness because novel mutations that caused psychiatric illness in some individuals nevertheless conferred something beneficial to others. The idea is that the brain is still evolving, it is an on-going experiment, and that any upside in human cognitive benefit to some is likely to have a downside in costs to others. It is an old idea that keeps on returning to vogue and it has just been given a face-lift by two recent genetic studies that conclude that the side-effects of the rapid growth of the human brain can be seen in psychiatric conditions like schizophrenia and bipolar illness (sometimes called manic depression).

Full-blown florid manic depressives are profoundly ill people. Their mood swings between psychotic mania and profound depression. It is estimated that 20% of manic depressives risk suicide if left untreated. Both manic depression and schizophrenia (which overlap, so far as symptoms and, probably, genetics is concerned) are considered dysgenic—sufferers tend to have far fewer children than normal individuals. The gene(s) causing these two illnesses should, therefore, tend to die out—but they don't. Manic depression has existed with a frequency of approximately 1% in all human populations for as long as it has been measured. This has led to the idea that, although the genes for manic depression cause serious illness in some individuals, they cause compensatory advantages in related individuals who share all or some complement of the genes. For instance, I met a New Yorker, Evelyn Gilman, who was constructing a family tree heavily 'infected' with mental illness on her husband's side. His father committed suicide, his brother had a life history of clinical depression, and the rest of the tree was littered with aunts and cousins who had some kind of affective disorder or another. The Gilmans' daughter, Barbara, had been a promising artist, then began to show signs of mania. She used to paint all night long, fuelling herself on bags of granulated sugar, swallowed in mouthfuls. She finally ran riot through her art school campus, suffering from bizarre paranoid delusions, and had to be hospitalized for good. Yet her father, Evelyn's husband, Herb, bucked the trend. He was a breezy, energetic entrepreneur who had built up from scratch one of the biggest chains of department stores in America. He had never lost a day to anything. Was he one of the 'compensated' individuals? Did those genes, in him, give him his boundless energy and optimism, and creative business drive? Is the old adage true—that genius is close to madness?

Kay Redfield Jamison is a manic depressive and currently a professor of psychiatry at Johns Hopkins University in the US. She wrote *Touched With Fire: Manic Depressive Illness and the Artistic Temperament*, which charts the occurrence of bipolar illness in creative and high-achieving families. When she took up a temporary post in London, in 1986, she sent out a questionnaire to hundreds of Britain's more prominent playwrights, artists, Royal Academicians, winners of the Booker Prize, and so on. All were asked to list details of any history of manic depression. The frequency of people actually having sought and received treatment for depression or manic depressive illness was 38%—against a background rate of 1% for the general population. The highest rate was amongst the poets—half of whom had sought professional help for their illness.

Robert Lowell, arguably the greatest American poet of the twentieth century, was often violently ill and frequently hospitalized. He called his illness 'a flaw in the motor' and 'dust in the blood'. His only relief was when his emotional pendulum passed through a period of hypomania on its way from peak to trough. He called it his 'magical orange grove in the midst of a nightmare' and it is when his mind raced and he poured out all his work. Lists of composers and political leaders tell the same story. There is something about the human qualities we treasure—inspirational leadership, literary, oratorical, and artistic fluency, boundless optimism, and a cognitive dissonance that leads to an energetic perseverance against all odds (all hypomanic traits), that seems connected with psychiatric genetics. An American researcher got a large number of America's chief political figures and captains of industry to respond to a questionnaire designed to reveal hypomanic traits. He concluded that they all rated so highly on a scale of hypomania that most of them would have been hospitalized by psychiatrists had they ever sought medical advice. Of course, none of them had; they were simply as high as kites, busy running the world!

Gordon Claridge, Daniel Nettles, and Tim Crow have all associated the mild side of schizophrenia, called schizotypal cognition, with creativity and divergent thinking while, as for manic depression, artists, poets, writers, and mathematicians are among the creative categories where schizophrenia is over-represented, compared to the general population. Vincent Van Gogh, Albert Einstein, Emily Dickinson, and Isaac Newton were all sufferers. Einstein's son developed full-blown schizophrenia, as did James Joyce's daughter. The *Beautiful Mind* mathematician, John Nash, was a florid schizophrenic but, in his saner moments, made extraordinary contributions to game theory.

Now Steve Dorus (formerly a research collaborator with Bruce Lahn, now at the University of Bath) has teamed up with evolutionary biologist Bernard Crespi, to take a modern stab at the relationship between genius and madness. If these supposed compensatory cognitive traits have been a feature of human cognitive evolution they must have helped to make us cognitively advanced, even unique. If so, many of the genes suspected over many years to be implicated in schizophrenia might actually have a history of positive selection unique to the human line. This is exactly what Crespi and Dorus set out to test.

They went into the literature on the genetics of schizophrenia and built up a list of 76 candidate genes that, at one time or another, had been implicated in the disorder. They used a number of statistical techniques to look for evidence that any of these genes had been subject to recent positive selection. Out of 76 genes, 14 stood out. The star performer was a gene named *DTNBP1*, which had a particularly strong signal for selection and is one of the genes with the strongest evidence of association with schizophrenia. On top of that, they found a further 12 genes which showed milder evidence of selection.

A pattern of ancient and modern evolution emerged. Four genes appeared to have been selected for since the split from the chimpanzee–human common ancestor, while two genes, *NRG* (neuroregulin)1 and *DISC* (disrupted in schizophrenia)1 had been very strongly selected for in both the human–chimpanzee lineages and, even earlier still, in the primate-origin lineages.

Crespi and Dorus also found that several other research groups trawling the human genome for evidence of selection and rapid evolution had turned up a further 16 genes that have clear links with schizophrenia. They include our old friends FOXP2 and Microcephalin, and the two genes we discussed in chapter 11, MAO-A and SLC6A4, that work in association with each other and are involved in personality, depression, mood, and thought disorders, antisocial conduct, novelty seeking, and substance abuse. Some of these genes behave in an aberrant way in dorsolateral, prefrontal and orbitofrontal cortices of schizophrenics. These are the very areas which govern our social cognition and help us plan ahead and behave appropriately in society. No wonder that disordered gene expression leads to disordered thought processes!

Crespi and Dorus have produced persuasive material that links recent evolutionary changes in neuro-anatomy and genes to mental illness on the one hand and creative mental processes in the normal population on the other. Many of the genes they have found are not specific to schizophrenia, but are involved in a host of brain-building, brain-wiring, and nerve-transmission processes. That such an impressive list of schizophrenia-related genes appear to have been so strongly selected for, both in earlier primate history, and particularly on the lineage specific to us humans, really does suggest that schizophrenia, specifically, is an illness with its roots going back 25 million years, when primate brains began to grow and organize fresh complexity. But it also has a more modern dimension that is peculiar to the human lineage since the split from the common ancestor. Schizophrenia, in terms of its building blocks in primate brain evolution, is something we share with other primates and yet is something which sets us apart.

Philipp Khaitovich, from the Max-Planck Institute in Leipzig, and an international team of colleagues have produced complementary research linking schizophrenia with human brain evolution. Specifically, they find that genes that have increased their expression and metabolites that have increased their concentrations in the brains of schizophrenics are both strongly related to energy metabolism and energy-expensive brain functions. Precisely the same gene expression and metabolite concentration changes appear to have been strongly positively selected during human evolution—they have altered in the human, but not the chimpanzee, lineage. So, schizophrenia seems synonymous with brain expansion.

The metabolites that changed their concentrations in the brain during human evolution, they say, are involved in the most energy-demanding processes in the human brain—maintenance of the electrical potential across cell membranes, vital for firing of neurons, and the continual synthesis of neurotransmitters. Human brain expansion meant more neurons firing rapidly over greater distances as nerve impulses integrated cognition between brain modules that were now further apart than their equivalent structures in the chimpanzee brain. This would have resulted in an increase in the length and diameter of neural connections and the number of synapses. The bigger and more complex the brain's wiring, the more energy needed to pump nerve impulses around it. The human brain, they reckon, must now be running very close to its metabolic limits and any upset in energy metabolism could cause cognitive dysfunction. In that respect, they note that schizophrenia is associated with structural and functional abnormalities in the fronto-temporal and fronto-parietal circuits which are connected by some of the longest-ranging projection neurons in the brain. They fire off nerve impulses at a high rate—requiring a great deal of energy. Evolution has played, and probably still is playing, with madness, and mental illness may be the price many of us continue to pay for evolution's continual tinkering with the human brain.

Brain evolution like this has produced a very singular animal, in turns creative, intelligent, restless, delusional, conciliatory, and aggressive. Self-domestication has undoubtedly permitted the brain-growth periods in gestation and prolonged childhood that have given rise to these unique forms of human cognition, and, in the process, very recently changed a large proportion of our genome versus that of the chimpanzee. But, according to some observers, it may also prove to be our downfall.

The late Canadian environmental scientist John A. Livingston, in his book *Rogue Primate*, pointed out that we are the only species of primate to self-domesticate and that we then went on to domesticate a wide range of plants and animals. Our 'arrested ontogeny'—the explosion of neotenous traits we discussed in chapter 11—occurred during this sustained period of self-domestication. That process led to our big brains, advanced intellects, peculiar and unique form of social cognition—our ability to be self-aware and to project into the mental life of others. But, for Livingston, it led consequentially to our reliance on ideas and ideology and our continual detachment from the State of Nature inhabited by the rest of the animal kingdom. We subjugated Nature—tamed it—domesticated it. It is why we are wrecking the planet. In Livingston's view our expanded cerebral cortex is the dysfunctional equivalent of a brain tumour and the higher-order cognition that has led to great music and literature, burgeoning science and technology, and powerful abstract thought processes on the one hand, has also led us to what he calls 'zero-order humanism' which is the assumed primacy of human interests over any other life-form:

> The human enterprise is necessary. The human future is imperative. The subsidy Earth and earthly non-human beings are required to pay for human immortality is, of course, factored out, not in. The realization of the future—the necessary advancement of the human monoculture—is to be sustained by Nature. The clear assumption is that Nature owes us. It is Nature's appointed task—its reason for being—to maintain and nourish the human project.

Livingston believed we would not stop until we had domesticated the whole planet—that we were the ultimate rogue species. He agreed with Jared Diamond that the invention of agriculture was one of the silliest things we, as a species, ever did. A remorseless ball started rolling in the Fertile Crescent some 8,000 years ago as monoculture agriculture supplanted wild ecological diversity—and where human overcrowding began to build up. The moving wave front of humanity spread out, and continues to spread today, colonizing new virgin territories and making them subservient to our needs. In the same way that we emasculated Nature by turning wild species into dull, sensory-deprived, overcrowding-tolerant animals totally dependent on us, argues Livingston, we have become totally dependent on and enslaved by our ideology and technology. Today we are presiding over a massive reduction in species and ecological diversity—the last great extinction. We have produced a range of oxymoronic concepts like 'sustainable development' by which we mean the exploitation business as usual, and 'wilderness areas', 'areas of outstanding natural beauty', and 'national parks', which are all for our benefit, not Nature's, and are perpetually under threat through the vested interests of 'development'.

Along the way, Livingston argued, we created power-based social institutions of a strongly hierarchical nature, powers that demanded compliance. For Livingston, the only way that human rights could arise from the concentration of power in these institutions was if they were willingly conveyed by or forcibly extracted from those who held the power. There is, he said, no such thing as a natural right. 'Power and politics and rights and duties', he said, 'are necessary human concoctions in order that our societies, twisted and deformed though they are, may function in some reasonable semblance of order and peacefulness.' We know from our history books what a violent, bloody, and protracted process this redistribution of power has been—one only has to look at women's suffrage and pro-choice, fights for independence from a succession of empires, equality for black Americans, and the fight against apartheid. Rights have always had to be fought for and perpetually defended against those who would take them away again.

Against this background let us briefly re-visit the case of Matthew Hiasl Pan, the chimp in Austria for whom campaigners are petitioning the Hague to have him granted human rights—to be acknowledged as a person. The main plank of the scientific argument in support of his case revolves around the claim that he should be re-classified as a human being because the genetics and cognition of chimps differs from humans by only a whisker.

For Hiasl Pan's supporters it is a matter of expediency. Were he to be granted human rights someone could be appointed his legal guardian—for clearly he is incompetent. But supposing Hiasl Pan, newly invested with humanness, decided to go on the rampage and emasculate his keeper—or legal guardian—or rip a baby girl's arm off? Could he—should he—be summarily shot, as happened to the two chimps who so savagely attacked St. James Davis in California? In what way could Hiasl Pan plead in a court of law? Would he be judged mentally incompetent to stand trial? Could it be said to be a crime of passion? A case of diminished responsibility? It's easy to see how farcical his status is. Rights are a specific construct and are to do with persons, not chimps. They come with duties and obligations, infringements of which can be penalized.

What we will have done, if we cave in and give Hiasl Pan, and any other great ape in trouble, human rights, will have been to spuriously admit a species into the human fold which is not as close to us genetically as we once thought it was, and not as close to us cognitively as we might have wished. Once we have done this it makes the case for human rights for other animal species more arbitrary—more arguable. Apart from the fact that we may have opened the floodgates into humanity, we will be in the preposterous situation of using the concept of human rights to effect beneficial outcomes for the rest of the animal (perhaps also the plant) kingdom. As John Livingston would have put it, we will have begun by admitting a wild animal—a chimpanzee—into a dysfunctional one-species club of domesticated planet-wreckers.

The philosopher Peter Singer argued that when we put human interests above the interests of any other species we are guilty of speciesism—a form of racism. When scientists, like Jane Goodall, agree to stand as expert witnesses in favour of human rights for Hiasl Pan, are they not unwittingly guilty of a similar type of mild racism because, by stressing the continuity between us and chimps, they make way for chimpanzees to join a club which always puts its interests first? Instead of spreading continuity they are simply broadening this species chauvinism. Chimpanzees exist in the State of Nature. John Livingston's answer to our deluded planet-wrecking is that we should embrace wildness and try and put ourselves, at least to some extent, back into the State of Nature. But once the genie is out of the bottle it is impossible to put it back. I don't think we have a chance of returning to Livingston's Garden of Eden and I, for one, would not like to be there. Like most of us, I suspect, I prefer to live in what he would have described as a delusional bubble in which technological optimism and faith in our unique universal human morality and sense of compassion pulls us and the planet through.

We are a truly exceptional primate with minds that are genuinely discontinuous to other animals. The fact that we have chosen to use the chimpanzee as an invaluable research tool to find out how we became human should not allow us to become over-familiar. We should learn to keep chimpanzees at arm's length—mentally and physically. Re-branding chimpanzees as humans will not save them from extinction. Rather, it behoves us, as humans, to find ways of managing our affairs that are far less ruthless and dismissive of the survival of the rest of the animal kingdom, and the environments on which they depend. Chimpanzees, the other great apes, the rest of the primates, and countless other animal and plant species are, today, quite literally perched on the brink of extinction because of us. We have to force ourselves to embrace the role of custodians of this planet and all the animal and plant genera upon it, chimpanzees included, regardless of their genetic proximity to us—which should be irrelevant. Otherwise we will be the first domesticated animal species which, as a result of its rapacious self-interest, will have driven the lot of us to extinction, leaving a bland and exhausted planet behind us.

GLOSSARY

base-pair Two nucleotides on opposite complementary strands of DNA connected by hydrogen bonds: adenine pairs with thymine and guanine with cytosine.

demographics The study of aspects of populations like size, growth rates, age structure, fertility, mortality, and migration, that can lead to population changes and affect the frequency of genes within populations.

exon A sequence of DNA in a gene that is transcribed into the corresponding RNA to form the template for the assembly of amino-acids into chains of protein.

fixation A situation in populations of animal or plant species where only one version or allele of a particular gene or locus is present because selection has favoured it over all other alleles to the extent that it has reached a frequency of 100%, i.e. all individuals have it.

gene expression The level of activity of a gene in coding for protein.

genetic markers Stretches of known, recognizable DNA sequence, which are spaced out along the chromosomes at intervals. It is possible to map the occasions when a particular marker and the affected condition turn up together in the same individual when the gene for the condition is located close enough to the marker for them both to be carried into the next generation on the same chromosome, surviving any recombination, or crossing-over, that occurs between the chromosomes during sexual reproduction. This is called linkage analysis.

genomics The study of all the nucleotide sequences in the chromosomes of an organism.

genotyping The use of biological techniques like polymerase chain reaction and DNA sequencing to determine parts or the whole of an organism's genotype, or genetic make-up.

grey matter Greyish-coloured tissue of the brain and spinal cord, composed of nerve cell bodies and their connections; as opposed to white matter, which is composed of myelinated, or sheathed, nerve fibres.

hominids A taxonomic family which includes humans, and their ancestors, and the other three other great apes; chimpanzees, gorillas and orang-utans.

hominins All the bipedal apes: present-day humans, and fossilized human ancestors, dating back to the split from the common ancestor approximately 6 million years ago.

hominoids A broader grouping of primates to include humans, the great apes, siamangs and gibbons.

indels Insertions or deletions of lengths of DNA sequence.

intron DNA sequence within a gene that is situated between the exons, or coding sequences. Introns are normally spliced away from the exons before the latter are copied into messenger RNA and thus code for amino-acid sequence in proteins.

Ka/Ks ratio Ratio of non-synonymous to synonymous mutations: the higher the ratio, the more likely that positive selection has operated on that particular mutation.

meiosis A process of cell division which comprises two nuclear divisions to form four cells, each containing exactly half the chromosomal complement of the original. These are the gametes—male and female—and when they recombine in sexual reproduction, normal diploid chromosomal number is restored.

mitosis The division of one cell into two daughter cells, involving a doubling, or duplication, of all chromosomes, such that the two resulting cells have the same genetic complement as the parent.

negative selection Sometimes referred to as purifying selection: the selective removal from the genome of alleles (gene variants containing one or more mutations resulting in nucleotide substitutions compared with the ancestral gene).

neocortex The outermost layer of the cerebral hemispheres involved in higher cognitive functions such as sensory perception, spatial reasoning. and conscious thought.

neuropil A highly complex brain tissue that lies outside the main nerve cells of the grey matter of the brain and comprises a huge network of fibres and synapse junctions coming from both neurons and astrocytes.

non-synonymous mutation A substitution of one of the bases of the DNA genetic code, which results in a change in the sequence of amino-acids in the protein that results from that gene's transcription.

nucleotide A sub-unit of DNA (deoxyribonucleic acid) or its complementary single-helix molecule, RNA (ribonucleic acid), which is composed of a nitrogenous base—adenine, guanine, thymine, or cytosine in DNA, with tyrosine substituted by uracil in RNA—together with a phosphate molecule and a sugar.

palaeontology The study of extinct and fossilised animals and plants.

phenotype The observable physical and biochemical characteristics of an organism as opposed to the genotype, which is the DNA sequence that gives rise to the phenotype via protein synthesis.

segmental duplication The duplication of a segment of DNA sequence that is more than 1,000 bases (1kb) in length.

selective sweep When positive selection propels an allele, or gene variant, to very high frequency in a population in such a way that other, more neutral variants of the same gene are eliminated.

sequence divergence The extent to which there has been substitution of one nucleotide for another within any given length of DNA.

synonymous mutation A change or substitution of one of the bases in the DNA code of a gene which does not cause a change in the corresponding amino-acid of the protein chain resulting from the transcription of that gene.

taxonomy A hierarchical classification system of organisms.

tectorial membrane A sheet of cells found in contact with the sensory hairs of the cochlea of the ear such that sound vibrations can be transmitted as nerve impulses.

ENDNOTES

Chapters 1–7

Throughout the chapters on genetics I refer to DNA nucleotides, nucleotide base-pairs, and substitution of bases in the form of point mutations. The double helix of DNA is composed of two long connected chains of nucleotides. There are only four DNA nucleotides: adenine, thymine, cytosine, and guanine. They are bonded together by hydrogen bonds such that adenine only binds to thymine and cytosine to guanine. DNA is thus a long double chain of A-T and C-G nucleotide pairs which are known as base-pairs. When the double strand of DNA is transcribed into single-strand messenger RNA, as a prelude to making protein, the thymine is replaced by uracil. Each triplet of bases in messenger RNA codes for a particular amino-acid. The long chain of a typical protein molecule is composed of many amino-acids. Occasionally, due to a point mutation, one of the DNA nucleotides gets substituted for another. This changes the composition of the triplets in the resulting messenger RNA. However, because there is some redundancy in the genetic code for each amino-acid it is possible for some DNA/RNA nucleotides to change without causing a substitution of the resulting amino-acid. In this case the mutation is silent—it is as if nothing has happened and the substitution is said to be synonymous. Occasionally, however, the substitution of one nucleotide in the DNA of a gene (and thence the messenger RNA) will cause the substitution of an amino-acid in the resulting protein chain. This may affect the way this protein behaves, with resulting knock-on effects for the organism. The vast majority of these changes are deleterious and are immediately weeded out by natural selection. However, many are neutral, which means they just quietly build up in the genome, or positive, in the sense that they will be selected for. These substitutions are therefore said to be non-synonymous.

For internet sources on DNA structure, RNA structure, and protein synthesis I recommend: <http://io.uwinnipeg.ca/~simmons/protsyn/sld001.htm> and following sequence of slides. Also helpful is <http://en.wikipedia.org/wiki/Point-mutation>.

Chapter 3. Brain-builders

There are a good many other genes, besides FOXP2, ASPM, and Microcephalin, where research in psychiatric genetics has led to interesting clues to human cognitive evolution. Christopher Walsh's lab at Harvard is investigating a number of other examples, including AHI1 and GPR56, where certain mutations cause Joubert's Syndrome and pathology of the frontal lobes respectively. Bruce Lahn's lab in Chicago has also been investigating a number of candidate genes underlying unique human cognitive development. Readers wanting to explore this area more fully are recommended to visit these labs' websites to learn more:

http://www.walshlab.org/re_about.php

http://www.genes.chicago.edu/lahn.html

Chapter 4. The Riddle of the 1.6%

In this book I have concentrated on gene expression differences in the brains of chimpanzees and humans. But there is also valuable work on gene expression in the liver by Yoav Gilad that also supports King's and Wilson's theory. For a full bibliography visit Gilad's website at:

http://www.genes.chicago.edu/gilad.html

Chapter 8. Povinelli's Gauntlet

Although I have cast this chapter almost exclusively around the research contretemps between Daniel Povinelli's Cognitive Evolution Group, and the Department of Developmental and Comparative Psychology at the Max Planck Institute for Evolutionary Anthropology at Leipzig, all contemporary human–chimpanzee comparative psychology owes a great deal to, and draws a great deal of inspiration from, the pioneering work of Wolfgang Kohler at his small research station on Tenerife in the Canary Islands, upon which he was marooned for the duration of the First World War. His experimental designs have formed the foundation for many of the experiments conducted in Louisiana, Leipzig, and elsewhere, and have also informed the comparative experiments with corvids. Readers are warmly recommended to read Kohler's own account of his research in *The Mentality of Apes* (Vintage Books, New York, 1927).

For further exhaustive detail of the contrasting research perspectives of the two main groups in this chapter readers are directed to their respective websites:

http://www.cognitiveevolutiongroup.org

http://www.eva.mpg.de/psycho/index.html

Finally, readers will, I hope, realize that I have restricted my discussion of chimp–human comparative cognitive psychology to the two groups in Louisiana and Leipzig for legitimate story-telling reasons. There are many other research groups around the world doing valuable research in this area. The work of the Scottish Primate Research Group, centred around the Universities of St Andrews, Stirling, and Edinburgh, is particularly outstanding. For details:

http://www.st-andrews.ac.uk/psychology/research/sprg/

Chapter 10. Inside the Brain—The Devil is in the Detail

We are indebted to Dr Leslie Brothers, an associate professor in psychiatry and behavioural sciences at the UCLA School of Medicine, for coining the term 'The Social Brain'. A full account of her work in this area can be found in her book *Friday's Footprint* (Oxford University Press, 1997).

For an introduction into the controversy as to whether the human mirror neuron system actually exists, or whether the indirect methods that claim to demonstrate it actually do so, I suggest 'No evidence of Human Mirror Neurons' by Olivier Morin, which appears on the International Cognition and Culture Institute website at <http://www.cognitionandculture.net/index.php?option=com_content&view=article& id=223:do-we-have-mirror-neurons-at-all&catid=9:neuro-dash&Itemid=34>.

BIBLIOGRAPHY

Chapter 1. From Distant Cousins to Close Family

'Wife: Mauled Man Tried Reasoning With Chimps During Attack.' NBC 4. 6 March 2005.

'How a chimp became part of the family: Man remains in coma after brutal attack.' Amy Argetsinger, *The Washington Post*. 26 May 2005.

'Nascar Veteran St. James Davis Continues To Astound Medical Officials With A Slow But Steady Progress.' Dave Grayson, *Racing West*. 5 July 2005.

'A Great Aping of Humans' Rights.' Josie Appleton, *Spiked online*. 8 June 2006.

'Spanish Parliament Approves "Human Rights" for Apes.' Lee Glendinning, *Guardian.co.uk*. 26 June 2008.

'Court to Rule if Chimp has Human Rights.' Kate Connolly, *The Observer*. 1 April 2007.

'Primate Rights?' Editorial, *Nature Neuroscience*, Vol. 10, Number 6. June 2007.

'Chimp Denied a Legal Guardian.' Ned Stafford, *Nature*. <http://www.nature.com/news/2007/070423/full/070423-9.html>.

Chimps: So Like us. (Film) Jane Goodall, Jane Goodall Institute. 2006.

'Court Won't Declare Chimp a Person.' William J Kole, *The Associated Press*. 27 September 2007.

'Look Into My Eyes.' James Mollison photography, *The Guardian Weekend*. 9 October 2004.

'The Chimpanzee Genome.' *Nature* special edition. 1 September 2005.

'Genomic Divergences between Humans and Other Hominoids and the Effective Population Size of the Common Ancestor of Humans and Chimpanzees.' Feng-Chi Chen and Wen-Hsiung Li, *American Journal of Human Genetics*, Vol. 68 (2001), pp. 444–56.

'Genomewide Comparison of DNA Sequences between Humans and Chimpanzees.' Ingo Ebersberger et al., *American Journal of Human Genetics*, Vol. 70 (2002), pp. 1490–7.

'Epilogue: A Personal Account of the Origins of a New Paradigm.' Morris Goodman, *Molecular Phylogenetics & Evolution*, Vol. 5 (February 1996), pp. 269–85.

'Primate Genomics: The Search for Genic Changes that Shaped being Human.' Morris Goodman, University of Chicago online report.

'Toward a Phylogenetic Classification of Primates Based on DNA Evidence Complemented by Fossil Evidence.' Morris Goodman et al., *Molecular Phylogenetics & Evolution*, Vol. 9 (June 1998), pp. 585–98.

'Implications of Natural Selection in Shaping 99.4% Nonsynonymous DNA Identity between Humans and Chimpanzees: Enlarging genus Homo.' Derek E. Wildman et al., *Proceedings of the National Academy of Sciences*, Vol. 100 (10 June 2003), pp. 7181–8.

Chapter 2. The Language Gene That Wasn't

The Language Instinct: The New Science of Language and Mind. Steven Pinker (Penguin, 1994).

'Feature-blind Grammar and Dysphasia.' M. Gopnik, *Nature*, Vol. 344 (19 April 1990), p 715.

'An Extended Family with a Dominantly Inherited Speech Disorder.' J. A. Hurst et al., *Developmental Medicine and Child Neurology*, Vol. 32 (1990), pp. 347–55.

'Praxic and Nonverbal Cognitive Deficits in a Large Family with a Genetically Transmitted Speech and Language Disorder.' Faraneh Vargha-Khadem et al., *Proceedings of the National Academy of Sciences*, Vol. 92 (January 1995), pp. 930–3.

'Neural Basis of an Inherited Speech and Language Disorder.' F. Vargha-Khadem et al., *Proceedings of the National Academy of Sciences*, Vol. 95 (13 October 1998), pp. 12695–700.

'The SPCH1 Region on Human 7q31: Genomic Characterization of the Critical Interval and Localization of Translocations Associated with Speech and Language Disorder.' Cecilia S. L. Lai et al., *American Journal of Human Genetics*, Vol. 67 (2000), pp. 357–68.

'Behavioural Analysis of an Inherited Speech and Language Disorder: Comparison with Acquired Aphasia.' K. E. Watkins et al., *Brain*, Vol. 125 (March 2002), pp. 452–64.

'MRI Analysis of an Inherited Speech and Language Disorder: Structural Brain Abnormalities.' K. E. Watkins et al., *Brain*, Vol. 125 (March 2002), pp. 465–78.

'Localisation of a Gene Implicated in a Severe Speech Language Disorder.' Simon E. Fisher et al., *Nature Genetics*, Vol. 18 (1998), pp. 168–70.

'FOXP2 in Focus: What Can Genes Tell Us About Speech and Language?' Gary F. Marcus and Simon E. Fisher, *Trends in Cognitive Sciences*, Vol. 7 (6 June 2003), pp. 257–62.

'Molecular Evolution of FOXP2, a Gene Involved in Speech and Language.' Wolfgang Enard et al., *Nature*, Vol. 418 (22 August 2002), pp. 869–72.

'Dissection of Molecular Mechanisms Underlying Speech and Language Disorders.' Simon E. Fisher, adapted from Keynote presentation at 'The Relationship of Genes, Environments, and Developmental Language Disorders' conference, Arizona, 2003.

'FOXP2 Expression During Brain Development Coincides with Adult Sites of Pathology in a Severe Speech and Language Disorder.' Cecilia S. L. Lai et al., *Brain*, Vol. 126 (22 July 2003), pp. 2455–62.

'FOXP2 and the Neuroanatomy of Speech and Language.' Faraneh Vargha-Khadem et al., *Nature Reviews*, Vol. 6 (February 2005), pp. 131–8.

'Genetic Components of Vocal Learning.' Constance Scharff and Stephanie A. White, *Annals New York Academy Sciences*, Vol. 1016 (2004), pp. 325–47.

'Scientists Find Parallels Between Human Speech and Bird Song Which Give Clues to Human Speech Disorders.' *Medical Research News*. 30 March 2004.

'The FOXP2 Gene, Human Cognition and Language.' Philip Lieberman, in *Integrative Approaches to Human Health and Evolution*. Proceedings of the International Symposium 'Integrative Approaches to Human Health and Evolution' held in Madrid, Spain, between 18 and 20 April 2005.

'Accelerated FOXP2 Evolution in Echolocating Bats.' Gang Li et al., *PLoS One*, Issue 9 (September 2007).

'Altered ultrasonic vocalization in mice with a disruption in the FoxP2 gene.' Weiguo Shu et al., *Proceedings of the National Academy of Sciences*, Vol. 102 (5 July 2005), pp. 9643–8.

Chapter 3. Brain-builders

'Primary Autosomal Recessive Microcephaly (MCPH1) Maps to Chromosome 8p22-pter.' Andrew P. Jackson et al., *American Journal of Human Genetics*, Vol. 63 (1998), pp. 541–6.

'Identification of *Microcephalin*, a Protein Implicated in Determining the Size of the Brain.' Andrew P. Jackson et al., *American Journal of Human Genetics*, Vol. 71 (2002), pp. 136–42.

'ASPM is a Major Determinant of Cerebral Cortical Size.' Jacquelyn Bond et al., *Nature Genetics*, Vol. 32 (22 September 2002), pp. 316–20.

'Aspm Specifically Maintains Symmetric Proliferative Divisions of Neuroepithelial Cells.' Jennifer L. Fish et al., *Proceedings of the National Academy of Sciences*, Vol. 103 (5 July 2006), pp. 10438–43.

'Accelerated Evolution of the ASPM Gene Controlling Brain Size Begins Prior to Human Brain Expansion.' Natalay Kouprina et al., *PLoS Biology*, Vol. 2, Issue 5 (May 2004), pp. 653–63.

'Evolution of the Human ASPM Gene, a Major Determinant of Brain Size.' Jianzhi Zhang, *Genetics*, Vol. 165 (December 2003), pp. 2063–70.

'*Microcephalin*, a Gene Regulating Brain Size, Continues to Evolve Adaptively in Humans.' Patrick D. Evans et al., *Science*, Vol. 309 (9 September 2005), pp. 1717–20.

'Ongoing Adaptive Evolution of ASPM, a Brain Size Determinant in Homo sapiens.' Nitzan Mekel-Bobrov et al., *Science*, Vol. 309 (9 September 2005), pp. 1720–2.

'Normal Variants of *Microcephalin* and ASPM Do Not Account for Brain Size Variability.' Roger P. Woods et al., *Human Molecular Genetics*, Vol. 15 (2006), pp. 2025–9.

'Comment on "Ongoing Adaptive Evolution of ASPM, a Brain Size Determinant in Homo sapiens".' Fuli Yu et al., *Science*, Vol. 316 (20 April 2007).

'Response to Comments by Timpson et al and Yu et al.' Nitzan Mekel-Bobrov and Bruce T Lahn, *Science*, Vol. 317 (24 August 2007), pp. 1036a + b.

'No Evidence that Polymorphisms of Brain Regulator Genes *Microcephalin* and *ASPM* are Associated with General Mental Ability, Head Circumference or Altruism.' J. Philippe Rushton et al., *Biology Letters*, 3(2) (22 April 2007), pp. 157–60. doi:10.1098/rsbl.2006.0586

'The Ongoing Adaptive Evolution of *ASPM* and *Microcephalin* Is Not Explained By Increased Intelligence.' Nitzan Mekel-Bobrov et al., *Human Molecular Genetics*, published online 12 January 2007.

'Bruce Lahn Profile: Links Between Brain Genes, Evolution, and Cognition Challenged.' Michael Balter, *Science*, 22 December, 2006, p. 1872.

'Head Examined: Scientist's Study Of Brain Genes Sparks a Backlash.' Antonio Regalado, *The Wall Street Journal*, 16 June 2006.

'Linguistic Tone is Related to the Population Frequency of the Adaptive Haplogroups of Two Brain Size Genes, *ASPM* and *Microcephalin*.' Dan Dediu and D. Robert Ladd, *Proceedings of the National Academy of Sciences* Early Edition, 30 May 2007. <http://www.pnas.org/cgi/doi/10.1073/pnas.0610848104>.

'Comparing the Human and Chimpanzee Genomes: Searching for Needles in a Haystack.' Ajit Varki and Tasha K. Altheide, *Genome Research* (2005), pp. 1746–58.

Chapter 4. The Riddle of the 1.6%

'Comparative Genetics: Which of Our Genes Make Us Human?' Ann Gibbons, *Science* (4 September 1998), pp. 1432–4.

'Positive Selection on the Human Genome.' Eric J. Vallender and Bruce T. Lahn, *Human Molecular Genetics*, Vol. 13 (2004), pp. 245–54.

'Accelerated Evolution of Nervous System Genes in the Origin of Homo Sapiens.' Steve Dorus et al., *Cell*, Vol. 119 (29 December 2004), pp. 1027–40.

'Natural Selection on Protein-Coding Genes in the Human Genome.' Carlos Bustamente et al., *Nature*, Vol. 437 (20 October 2005), pp. 1153–7.

'Human Cognitive Abilities Resulted from Intense Evolutionary Selection, Says Lahn.' Catherine Gianaro, *University of Chicago Chronicle*, 6 January 2005.

'Eighty Percent of Proteins are Different Between Humans and Chimpanzees.' Galina Glazko et al., *Gene*, Vol. 346 (2005), pp. 215–19.

'Evolution at Two Levels in Humans and Chimpanzees.' Mary-Claire King and A. C. Wilson, *Science*, Vol. 188, No. 4184 (11 April 1975), pp. 107–16.

'A New (Mis)take on an Old Paper.' Mike Dunford, *ScienceBlogs*, 2 July 2007.

'Ancient and Recent Positive Selection Transformed Opioid cis-Regulation in Humans.' Matthew V. Rockman et al., *PLoS Biology*, Vol. 3 (December 2005), pp. 1–12.

'Promoter Regions of Many Neural- and Nutrition-Related Genes Have Experienced Positive Selection During Human Evolution.' Ralph Haygood et al., *Nature Genetics*, Vol. 39, Number 9 (September 2007), pp. 1140–4.

'The 1% Solution.' Constance Holden, *ScienceNOW Daily News*, 13 August 2007.

'Human Brain Evolution: Insights from Microarrays.' Todd M. Preuss et al., *Nature Reviews Genetics*, Vol. 5 (November 2004), pp. 850–60.

'Elevated Gene Expression Levels Distinguish Human From Non-Human Primate Brains.' Mario Caceres et al., *Proceedings of the National Academy of Sciences*, Vol. 100, No. 22 (October 2003), pp. 13030–5.

'Intra- and Interspecific Variation in Primate Gene Expression Patterns.' Wolfgang Enard et al., *Science*, Vol. 296 (12 April, 2002), pp. 340–3.

'Evolution of Primate Gene Expression.' *Nature Reviews Genetics*, Vol. 7 (September 2006), pp. 693–702.

'What Separates Man From Chimp?' *myDNA News*, 23 March 2006.

'Increased Cortical Expression of Two Synaptogenic Thrombospondins in Human Brain Evolution.' Mario Caceres et al., *Cerebral Cortex*, 20 December 2006.

'Brain Evolution Studies Go Micro.' Michael Balter, *Science*, Vol. 315 (2 March 2007), pp. 1208–11.

Chapter 5. Less Is More

'Sequencing The Chimpanzee Genome: Insights Into Human Evolution And Disease.' Maynard V. Olsen and Ajit Varki, *Nature Reviews Genetics*, Vol. 4 (January 2003), pp. 20–8.

'A Structural Difference Between the Cell Surfaces of Humans and the Great Apes.' Elaine Muchmore et al., *American Journal of Physical Anthropology*, Vol. 107 (1998), pp. 187–98.

'Loss of N-Glycolylneuraminic Acid in Human Evolution.' Els C. M. Brinkman-Van der Linden al., *Journal of Biological Chemistry*, Vol. 275 (24 March 2000), pp. 8633–40.

'Prehistory of falciparum malaria.' John Hawks Weblog, 9 June 2005.

'A Mutation in Human CMP-Sialic Acid Hydroxylase Occurred After the Homo-Pan Divergence.' Hsun-Hua Chou et al., *Proceedings of the National Academy of Sciences*, Vol. 95 (September 1998), pp. 11751–6.

'Loss of Siglec Expression on T Lymphocytes during Human Evolution.' Dzung Nguyen et al., *Proceedings of the National Academy of Sciences*, Vol. 103 (16 May 2006), pp. 7765–70.

'T Cell "Brakes" Lost During Human Evolution.' Debra Kain, *UCSD Press Release*, 1 May 2006.

'The Evolutionary History of the *CCR5-del* 32 HIV-Resistance Mutation.' Alison P. Galvani and John Novembre, *Microbes and Infection*, 7 (2005), pp. 302–9.

'The Geographic Spread of the *CCR5 - del. 32* HIV-Resistance Allele.' John Novembre, Alison P. Galvani, and Montgomery Slatkin, *PLoS Biology*, Vol. 3, Issue 11 (November 2005), p. 1.

'What Caused the Black Death?' C. J. Duncan and S. Scott, *Postgraduate Medical Journal* 81 (2005), pp. 315–20.

'Case Reopens on Black Death Cause.' Debora MacKenzie, *New Scientist online*, 11 September 2003.

'The Black Death and AIDS: *CCR5-del* 32 in Genetics and History.' S. K. Cohn and L. T. Weaver, *QJM: An International Journal of Medicine*, Vol. 99, Issue 8 (August 2006) pp. 497–503.

'Households and Plague in Early Modern Italy.' Samuel K. Cohn, Jr. and Guido Alfani, *Journal of Interdisciplinary History*, Autumn 2007, pp. 177–205.

'The Case for Selection at CCR5 - del. 32.' Pardis Sabeti et al., *PLoS Biology*, Vol. 3, Issue 11 (November 2005), p. 1963.

'Myosin Gene Mutation Correlates With Anatomical Changes in the Human Lineage.' Hansell H. Stedman et al., *Nature*, Vol. 428 (25 March 2004), pp. 415–18

'Human Genetics: Muscling In on Hominid Evolution.' Peter Currie, *Nature*, Vol. 428 (25 March 2004), News and Views, pp. 373–4.

'Gene Losses During Human Origins.' Xiaoxia Wang, Wendy E. Grus, and Jianzhi Zhang, *PLoS Biology*, Vol. 4, Issue 3 (March 2006), pp. 306–77.

Chapter 6. More Is Better

'Enhancing the Hominoid Brain.' Melissa Phillips, *The Scientist*, 20 September 2004.

'Birth and Adaptive Evolution of a Hominoid Gene that Supports High Neurotransmitter Flux.' Fabien Burki and Henrik Kaessmann, *Nature Genetics*, Vol. 36 (October 2004), pp. 1061–3.

'The Jewels of our Genome: The Search for the Genomic Changes Underlying the Evolutionarily Unique Capacities of the Human Brain.' James M. Sikela, *PLoS Genetics*, Vol. 2, Issue 5 (May 2006), pp. 646–55.

'Lineage-Specific Gene Duplication and Loss in Human and Great Ape Evolution.' Andrew Fortna et al., *PLoS Biology*, Vol. 2, Issue 7 (July 2004), pp. 937–54.

'Gene Duplications Give Clues to Humanness.' Jon Cohen, *ScienceNOW Daily News*, 30 July 2007.

'Human Lineage-Specific Amplification, Selection, and Neuronal Expression of DUF1220 Domains.' Magdalena C. Popesco et al., *Science*, Vol. 313 (1 September 2006), pp. 1304–7.

'Number of Copies of Immune-Response Gene Linked to HIV/AIDS Suscepti-bility.' Press Release, University of Texas Health Sciences Center, San Antonio, 7 January 2005.

'A Genome-Wide Comparison of Recent Chimpanzee and Human Seg-mental Duplications.' Ze Cheng et al., *Nature*, Vol. 437 (September 2005), pp. 88–93.

'Comparative Sequencing of Human and Chimpanzee MHC Class 1 Regions Unveils Insertions/Deletions as the Major Path to Genomic Divergence.' Tatsuya Anzai et al., *Proceedings of the National Academy of Sciences*, Vol. 100 (24 June 2003), pp. 7708–13.

'Recurrent Duplication-Driven Transposition of DNA During Hominoid Evo-lution.' Matthew E. Johnson et al., *Proceedings of the National Academy of Sciences*, Vol. 103, No. 47 (21 November 2006), pp. 17626–31.

'Primate Segmental Duplications: Crucibles of Evolution, Diversity and Disease.' Jeffrey A. Bailey and Evan E. Eichler, *Nature Reviews Genetics*, Vol. 7 (July 2006), pp. 552–64.

'Positive Selection of a Gene Family during the Emergence of Humans and African Apes.' Matthew E. Johnson et al., *Nature*, Vol. 413 (October 2001), pp. 514–18.

'Segmental Duplications and the Evolution of the Primate Genome.' Rhea Val-lente Samonte and Evan E. Eichler, *Nature Reviews Genetics*, Vol. 3 (January 2002), pp. 65–71.

'Genetic Breakthrough that Reveals the Differences between Humans.' Steve Connor, *The Independent*, 23 November 2006.

'Diet and the Evolution of Human Amylase Copy Number Variation.' George H. Perry, Nathaniel J. Dominy et al., *Nature Genetics*, Vol. 39 (9 September 2007), pp. 1256–60.

'Extra Gene Copies Were Enough to Make Early Humans' Mouths Water.' *Science Daily*, 9 September 2007.

'Savanna Chimpanzees Use Tools to Harvest the Underground Storage Organs of Plants.' R. Adriana Hernandez-Aguilar et al., *Proceedings of the National Academy of Sciences*, 21 November 2007.

'Study Finds Evidence of Genetic Response to Diet.' Nicholas Wade, *New York Times*, 10 September 2007.

'Accelerated Rate of Gene Gain and Loss in Primates.' Matthew W. Hahn et al., *Genetics*, Vol. 177 (November 2007), pp. 1941–9.

Chapter 7. Aladdin's Cave

'Genomic DNA Insertions and Deletions Occur Frequently Between Humans and Nonhuman Primates.' Kelly A. Frazer et al., *Genome Research*, 25 November 2002.

'Newly Discovered Gene May Hold Clues to Evolution of Human Brain Capacity.' *CBSE News for UC Santa Cruz*, 21 August 2006.

'Non-coding DNA Could Hold Secrets to What "Makes Us Human".' David Haussler, *Nature*, Vol. 443 (14 September 2006).

'An RNA Gene Expressed During Cortical Development Evolved Rapidly in Humans.' Katherine S. Pollard et al., *Nature*, Vol. 443 (September 2006), pp. 167–72.

'Forces Shaping the Fastest Evolving Regions in the Human Genome.' Katherine S. Pollard et al., *PLoS Genetics*, Vol. 2 (October 2006), pp. 1599–1611.

'Recent Progress in Understanding the Role of Reelin in Radial Neuronal Migration.' Eckart Förster et al., *European Journal of Neuroscience* Vol. 23 (2006), pp. 901–9.

'Flipped Genetic Sequences Illuminate Human Evolution And Disease.' *Science Daily*, 30 October 2005.

'Discovery of Human Inversion Polymorphisms by Comparative Analysis of Human and Chimpanzee DNA Sequence Assemblies.' Lars Feuk et al., *PLoS Genetics*, Vol. 1 (October 2005).

'Chromosomal Rearrangements and the Genomic Distribution of Gene-Expression Divergence in Humans and Chimpanzees.' Tomas Marques-Bonet et al., *Trends in Genetics*, Vol. 20, No. 11 (November 2004).

'Chromosomal Speciation and Molecular Divergence – Accelerated Evolution in Rearranged Chromosomes.' Arcadi Navarro and Nick H. Barton, *Science*, Vol. 300 (11 April 2003), pp. 321–4.

'A Process for Human/Chimpanzee Divergence.' Alec McAndrew, *Molecular Biology*, 3 May 2003.

'A Human-Specific Mutation Leads to the Origin of a Novel Splice Form of Neuropsin (KLK8), a Gene Involved in Learning and Memory.' Zhi-Xiang Lu et al., *Human Mutation*, 2007.

'Global Analysis of Alternative Splicing Differences between Humans and Chimpanzees.' John A. Calarco et al., *Genes & Development*, Vol. 21 (2007), pp. 2963–75.

'Alternative Splicing.' Jennifer Michalowski, *HHMI Bulletin*, September 2005.

Chapter 8. Povinelli's Gauntlet

Our Inner Ape. Frans De Waal (London: Granta Books, 2005).

'Why Humans are Superior to Apes.' Helene Guldberg, *Spiked online Essay*, 24 February 2004.

'Man Is More Than a Beast.' Helene Guldberg, *Spiked Science*, 3 November 2005.

Machiavellian Intelligence: Social Expertise and the Evolution of Intellect in Monkeys, Apes and Humans. A. Whiten and R. W. Byrne (Oxford University Press, 1988).

'Anecdotes, Training, Trapping and Triangulating: Do Animals Attribute Mental States?' C. M. Heyes, *Animal Behaviour*, Vol. 46 (1993), pp. 177–88.

'Reflections on Self-Recognition in Primates.' C. M. Heyes, *Animal Behaviour*, Vol. 47 (1994), pp. 909–19.

'The Chimpanzee's Mind: How Noble in Reason? How Absent of Ethics?' Daniel J. Povinelli and Laurie R. Godfrey, in *Evolutionary Ethics*, ed. M. Nitecki (Albany, NY: SUNY Press, 1993).

'Theory of Mind in Nonhuman Primates.' C. M. Heyes, *Behavioural and Brain Sciences*, Vol. 21 (1998), pp. 101–34.

Folk Physics for Apes. Daniel J. Povinelli (New York: Oxford University Press, 2000).

'Behind the Ape's Appearance: Escaping Anthropocentrism in the Study of Other Minds.' Daniel J. Povinelli, *Daedalus*, Vol. 133, No. 1 (Winter 2004), pp. 29–41.

'Chimpanzee Minds: Suspiciously Human?' Daniel J. Povinelli and Jennifer Vonk, *Trends in Cognitive Science*, Vol. 7 (April 2003), pp. 157–60.

'Do Chimpanzees Know What Conspecifics Know?' Brian Hare et al., *Animal Behaviour*, Vol. 61 (2001), pp. 139–51.

'Chimpanzees Versus Humans: It's Not That Simple.' Michael Tomasello, Josep Call, and Brian Hare, *Trends in Cognitive Sciences*, Vol. 7 (June 2003), pp. 239–40.

'Chimpanzees Understand Psychological States – The Question Is Which Ones and To What Extent.' Michael Tomasello, Josep Call, and Brian Hare, *Trends in Cognitive Sciences*, Vol. 7 (2003), pp. 153–6.

'We Don't Need a Microscope to Explore the Chimpanzee's Mind.' Daniel J. Povinelli and Jennifer Vonk, *Mind & Language*, Vol. 19 (February 2004), pp. 1–28.

'Chimpanzees Deceive a Human Competitor by Hiding.' Brian Hare et al., *Cognition*, Vol. 101 (2006), pp. 495–514.

'The Domestication of Social Cognition in Dogs.' Brian Hare et al., *Science*, Vol. 298 (22 November 2002), pp. 1634–6.

'Human-like Social Skills in Dogs?' Brian Hare and Michael Tomasello, *Trends in Cognitive Sciences*, Vol 9 (2005), pp. 439–44.

'A Simple Reason for a Big Difference: Wolves Do Not Look Back at Humans, but Dogs Do.' Adam Miklosi et al., *Current Biology*, Vol. 13 (29 April 2003), pp. 763–6.

'Comparative Social Cognition: What Can Dogs Teach Us?' A. Miklosi et al., *Animal Behaviour*, Vol. 67 (2004), pp. 995–1004.

'On the Lack of Evidence that Non-Human Animals Possess Anything Remotely Resembling a "Theory of Mind".' Derek C. Penn and Daniel J. Povinelli, *Philosophical Transactions of the Royal Society B* (2007), published online, doi:10.1098/rstb.2006.2023

'The Comparative Delusion: The "Behaviouristic"/"Mentalistic" Dichotomy in Comparative Theory of Mind Research.' Derek C. Penn and Daniel J. Povinelli, in *Oxford Handbook of Philosophy and Cognitive Science*, ed. R. Samuels and S. P. Stich (Oxford University Press, 2008).

Chapter 9. Clever Corvids

'Comparing the Complex Cognition of Birds and Primates.' Nathan J. Emery and Nicola S. Clayton, in *Comparative vertebrate cognition: are primates superior to non-primates?*, ed. L. J. Rogers and G. Kaplan (Kluwer Academic/Plenum Publishers, 2004).

'The Mentality of Crows: Convergent Evolution of Intelligence in Corvids and Apes.' Nathan J. Emery and Nicola S. Clayton, *Science*, Vol. 306 (10 December 2004), pp. 1903–7.

'Corvid Cognition.' Nicola Clayton and Nathan Emery, *Current Biology*, Vol. 15 (8 February 2005), pp. 80–81.

'Cognitive Ornithology: The Evolution of Avian Intelligence.' Nathan J. Emery, *Philosophical Transactions of the Royal Society B*, Vol. 361 (2006), pp. 23–43.

Crows. Candace Savage (Greystone Books, 2005).

In The Company of Crows and Ravens. John A. Marzluff and Tony Angell (Yale University Press, 2005).

'Food-Caching Western Scrub-Jays Keep Track of Who Was Watching When.' Joanna M. Dally et al., *Science*, Vol. 312 (16 June 2006), pp. 1662–5.

'It Takes a Thief to Know a Thief.' Nicola S. Clayton, University of Cambridge, Department of Experimental Psychology website.

'Planning for the Future by Western Scrub-jays.' C. R. Raby et al., *Nature*, Vol. 445 (22 February 2007), pp. 919–21.

'Investigating Physical Cognition in Rooks.' Amanda Seed et al., *Current Biology*, Vol. 16 (April 2004), pp. 697–701.

'Leading a Conspecific Away from Food in Ravens (Corvus corax)?' Thomas Bugnyar and Kurt Kotrschal, *Animal Cognition*, Vol. 7, No. 2 (April 2004), pp. 69–76.

'Ravens Differentiate between Knowledgeable and Ignorant Competitors.' Thomas Bugnyar and Bernd Heinrich, *Proceedings of the Royal Society B*, Vol. 272 (2005), pp. 1641–6.

'Direct Observations of Pandanus-Tool Manufacture and Use by a New Caledonian Crow.' Gavin R. Hunt and Russell D. Gray, *Animal Cognition*, Vol. 7 (2004), pp. 114–20.

'The Crafting of Hook Tools by Wild New Caledonian Crows.' Gavin R. Hunt and Russell D. Gray, *Proceedings of the Royal Society of London B*, Vol. 271, Biology Letters Suppl. 3 (2004), pp. S88–S90.

'The Right Tool for the Job: What Strategies Do Wild New Caledonian Crows Use?' Gavin R. Hunt et al., *Animal Cognition*, doi 10.1007/s10071-006-0047-2.

'Spontaneous Metatool Use by New Caledonian Crows.' Alex H. Taylor et al., *Current Biology*, Vol. 17 (4 September 2007), pp. 1504–7.

'Shaping of Hooks in New Caledonian Crows.' Alex A. S. Weir et al., *Science*, Vol. 297 (9 August 2002), p. 981.

Oxford University Behavioural Ecology Research Group website: <http://users.ox.ac.uk/~groups/tools/crow_natural_history.shtml>.

'A New Caledonian crow Creatively Re-designs Tools by Bending or Unbending Aluminium Strips.' Alex A. S. Weir and Alex Kacelnik, *Animal Cognition*, doi 10.1007/s10071-006-0052-5.

'Tool Selectivity in a Non-Primate, the New Caledonian Crow.' Jackie Chappell and Alex Kacelnik, *Animal Cognition*, Vol. 5 (2002), pp. 71–8.

'Selection of Tool Diameter by New Caledonian Crows.' Jackie Chappell and Alex Kacelnik, *Animal Cognition*, Vol. 7 (2004), pp. 121–7.

Chapter 10. Inside The Brain—The Devil is in the Detail

'What is it Like to Be a Human?' Todd M. Preuss, in *The Cognitive Neurosciences III*, 3rd edition, ed. M. S. Gazzaniga (Cambridge, MA: The MIT Press, 2005), pp. 5–22.

'Taking the Measure of Diversity: Comparative Alternatives to the Model-Animal Paradigm in Cortical Neuroscience.' Todd M. Preuss, *Brain, Behaviour and Evolution*, Vol. 55 (2000), pp. 287–99.

'The Brain and its Main Anatomical Subdivisions in Living Hominoids Using Magnetic Resonance Imaging.' Katerina Semendeferi and Hannah Damasio, *Journal of Human Evolution*, Vol. 38 (2000), pp. 317–32.

'Human and NonHuman Primate Brains: Are They Allometrically Scaled Versions of the Same Design?' James K. Rilling, *Evolutionary Anthropology*, Vol. 15 (2006), pp. 65–77.

'Brain of Chimpanzee Sheds Light on Mystery of Language.' Sandra Blakeslee, *New York Times*, 13 January 1998.

'Limbic Frontal Cortex in Hominoids: A Comparative Study of Area 13.' Katerina Semendeferi et al., *American Journal of Physical Anthropology*, Vol. 106 (1998), pp. 129–55.

'The Primate Neocortex in Comparative Perspective Using Magnetic Resonance Imaging.' James K. Rilling, *Journal of Human Evolution*, Vol. 37 (1999), pp. 191–223.

'Brain Circuitry Involved in Language Reveals Differences in Man, Non-Human Primates.' *Medical College of Georgia News*, 4 September 2001.

'The Evolution of the Arcuate Fasciculus Revealed with Comparative DTI.' James K. Rilling et al., *Nature Neuroscience*, advanced online publication 23 March, 2008, doi:10.1038/nn2072

'Prefrontal Cortex in Humans and Apes: A Comparative Study of Area 10.' Katerina Semendeferi et al., *American Journal of Physical Anthropology*, Vol. 114, Issue 3 (2001), pp. 224–41.

'A Comparative Volumetric Analysis of the Amygdaloid Complex and Basolateral Division in the Human and Ape Brain.' Nicole Barger et al., *American Journal of Physical Anthropology*, Vol. 134 (2007), pp. 392–403.

'A Comparison of Resting-State Brain Activity in Human and Chimpanzees.' James K. Rilling et al., *Proceedings of the National Academy of Sciences*, Vol. 104 (23 October 2007), pp. 17146–51.

'Are You Conscious of your Precuneus?' The Neurocritic blog, 25 June 2006.

'A Neuronal Morphologic Type Unique to Humans and Great Apes.' Esther A. Nimchinsky et al., *Proceedings of the National Academy of Sciences*, Vol. 96 (April 1999), pp. 5268–73.

'Two Phylogenetic Specializations in the Human Brain.' John Allman et al., *The Neuroscientist*, Vol. 8 (2002), pp. 335–46.

'Grasping the Intentions of Others with One's Own Mirror Neuron System.' Marco Iacoboni et al., *PLoS Biology*, Vol. 3 (March 2005), pp. 529–35.

Mirrors in the Brain. Giacomo Rizzolatti and Corrado Sinigaglia (Oxford University Press, 2007).

'The Mirror Neuron System and the Consequences of its Dysfunction.' Marco Iacoboni and Mirella Dapretto, *Nature Reviews neuroscience*, Vol. 7 (December 2006), pp. 942–51.

'The Mirror-Neuron System.' Giacomo Rizzolatti and Laila Craighero, *Annual Review of Neuroscience* Vol. 27 (2004), pp. 169–92.

'Cells That Read Minds.' Sandra Blakeslee, *The New York Times*, 10 January 2006.

'A Small Part of the Brain, and Its Profound Effects.' Sandra Blakeslee, *The New York Times*, 6 February 2007.

'Mirror Neurons and Imitation Learning as the Driving Force behind "the Great Leap Forward" in Human Evolution.' V. S. Ramachandran, *The Third Culture, Edge.* Org.No 69 (29 May 2000). [See also 'The Neurology of Self-Awareness', *Edge* 10th Anniversary Essay, 2007.]

'Understanding Others: Imitation, Language, Empathy.' Marco Iacoboni, in *Perspectives on Imitation: From Mirror Neurons to Memes*, ed. Susan Hurley and Nick Chater (Cambridge, MA: MIT Press, 2005).

'The Neural Basis of Mentalizing.' Chris D. Frith and Uta Frith, *Neuron*, Vol. 50 (18 May 2006), pp. 531–4.

'Brain Region Linked to Metaphor Comprehension.' *Science News*, 26 May 2006.

'The Social Brain.' Ralph Adolphs, *Engineering & Science* No. 1 (2006).

'Uniquely Human Social Cognition.' Rebecca Saxe, *Current Opinion in Neurobiology*, Vol. 16 (2006), pp. 235–9.

'Demystifying Social Cognition: a Hebbian Perspective.' Christian Keysers and David I. Perrett, *Trends in Cognitive Sciences*, Vol. 8 (November 2004).

'Both of us Disgusted in My Insula: The Common Neural Basis of Seeing and Feeling Disgust.' Bruno Wicker et al., *Neuron*, Vol. 40 (30 October 2003), pp. 655–64.

'Moral Intuition: Its Neural Substrates and Normative Significance.' James Woodward and John Allman, *Journal of Physiology (Paris)*, Vol. 101 (July 2007), pp. 179–202.

Moral Minds: How Nature Designed Our Universal Sense Of Right And Wrong. Marc D. Hauser (Ecco/HarperCollins, 2006).

Chapter 11. The Ape That Domesticated Itself

'Evidence of Very Recent Human Adaptation: Up to 10% of Human Genome May Have Changed.' *Science Daily*, 12 July 2007.

'Civilisation has Left its Mark on our Genes.' Bob Holmes, *New Scientist*, 19 December 2005.

'Global Landscape of Recent Inferred Darwinian Selection for Homo sapiens.' Eric T. Wang et al., *Proceedings of the National Academy of Sciences*, Vol. 103 (3 January 2006), pp. 135–40.

'A Map of Recent Positive Selection in the Human Genome.' Benjamin F. Voight et al., *PLoS Biology*, Vol. 4 (March 2006), pp. 446–58.

'Localizing Recent Adaptive Evolution in the Human Genome.' Scott H. Williamson et al., *PLoS Genetics*, Vol. 3 (June 2007), pp.901–15.

'Positive Selection in *NAT2* Gene.' Dienekes' Anthropology Blog, 22 December 2005.

'Blue-eyed Humans Have a Single, Common Ancestor.' *Eurekalert*, 30 January 2008.

'Recent Acceleration of Human Adaptive Evolution.' John Hawks et al., *Proceedings of the National Academy of Sciences* Early Edition, 2007, <http://www.pnas.org/cgi/doi/10.1073/pnas.0707650104>.

Growing Young. Ashley Montagu (Greenwood Publishing Group, 1988).

Ontogeny and Phylogeny. Steven J. Gould (Cambridge: Belknap Press, 1977).

From Non-Human to Human Mind: What Changed and Why? Brian Hare (Current Directions in Psychological Sciences, in press).

'Tolerance Allows Bonobos to Outperform Chimpanzees on a Cooperative Task.' Brian Hare et al., *Current Biology*, Vol. 17 (3 April 2007), pp. 619–23.

'From Wild Wolf to Domesticated Dog: Gene Expression Changes in the Brain.' Peter Saetre et al., *Molecular Brain Research*, Vol. 126 (2004), pp. 198–206.

Evolving Brains. John M Allman (Scientific American Library Series, No. 68, January 1999).

'Social Cognitive Evolution in Captive Foxes Is a Correlated By-Product of Experimental Domestication.' Brian Hare et al., *Current Biology*, Vol. 15 (8 February 2005), pp. 226–30.

'Russian Domesticated Foxes: A Status Report.' W. Tecumseh Fitch, personal publication, November 2002.

'Early Canid Domestication: The Farm-Fox Experiment.' Lyudmila N. Trut, *American Scientist*, Vol. 87 (March–April 1999), pp. 160–9.

DOGS: A Startling New Understanding of Canine Origin, Behaviour and Evolution. Ray and Lorna Coppinger (New York: Scribner, 2001).

Demonic Males: Apes and the Origins of Human Violence. Richard Wrangham and Dale Peterson (Boston: Houghton Mifflin Co., 1996).

'The Evolution of Cooking, A Talk with Richard Wrangham.' Edge.org, 28 February 2001.

'In Our Genes.' Henry Harpending and Gregory Cochran, *Proceedings of the National Academy of Sciences*, Vol. 99 (January 8, 2002), pp. 10–12.

'Evidence of Positive Selection Acting at the Human Dopamine Receptor D4 Gene Locus.' Yuan-Chun Ding et al., *Proceedings of the National Academy of Sciences*, Vol. 99 (January 8, 2002), pp. 309–14.

'Is ADHD An Advantage For Nomadic Tribesmen?' *Science Daily*, 10 June 2008.

'Microsatellite Instability Generates Diversity in Brain and Sociobehavioural Traits.' Elizabeth A. D. Hammock and Larry J. Young, *Science*, Vol. 308 (June 2005), pp. 1630–4.

'"Ruthlessness gene" discovered.' Michael Hopkin, *Naturenews*, published online, 4 April 2008.

'Are We Genetically Programmed to Be Generous? Israeli Scientists Say Yes.' Physorg.com, 6 December 2007.

'AVPR1a and SLC6A4 Gene Polymorphisms Are Associated with Creative Dance Performance.' Rachel Bachner-Melman et al., *PLoS Genetics*, Vol. 1 (September 2005), pp. 394–403.

'Genetic Variation in the Vasopressin Receptor 1a Gene (AVRP1A) Associates with Pair-Bonding Behaviour in Humans.' Hasse Walum et al., *Proceedings of the National Academy of Sciences*, Vol. 105 (16 September 2008), pp. 14153–6.

'Linking Emotion to the Social Brain.' Klaus-Peter Lesch, *EMBO reports*, Vol. 8, 2007.

'5HTT – Psychiatry's Big Bang.' McMan's Depression and Bipolar Web, <http://www.mcmanweb.com/>.

'Influence of Life Stress on Depression: Moderation by a Polymorphism in the 5-HTT Gene.' Avshalom Caspi et al., *Science*, Vol. 301 (18 July 2003), pp. 386–9.

'Role of Genotype in the Cycle of Violence in Maltreated Children.' Avshalom Caspi et al., *Science*, Vol. 297 (2 August 2002), pp. 851–4.

'Nature and Nurture Predispose to Violent Behaviour: Serotonergic Genes and Adverse Childhood Environment.' Andreas Reif et al., *Neuropsychopharmacology*, Vol. 32 (2007), pp. 2375–83.

'An Interview with Prof. K. P. Lesch.' in-cites, May 2007, <http://www.in-cites.com/papers/KPLesch.html>.

'Tracking the Evolutionary History of a "Warrior" Gene.' *Science*, Vol. 304 (7 May 2004), p. 814.

'Neural Mechanisms of Genetic Risk for Impulsivity and Violence in Humans.' Andreas Meyer-Lindenberg et al., *Proceedings of the National Academy of Sciences*, Vol. 103 (18 April 2006), pp. 6269–74.

'Why are there Shorts and Longs of the Serotonin Transporter Gene?' <http://www.psycheducation.org/mechanism/4WhyShortsLongs.htm>.

'5-HTTLPR Polymorphism Impacts Human Cingulate-Amygdala Interactions: A Genetic Susceptibility Mechanism for Depression.' Lukas Pezawas et al., *Nature Neuroscience*, Vol. 8 (2005), pp. 828–34.

Chapter 12. Chimps Aren't Us

'Adaptive Evolution of Genes Underlying Schizophrenia.' Bernard Crespi, Kyle Summers, and Steve Dorus, *Proceedings of the Royal Society B*, Vol. 274 (2007), pp. 2801–10.

'Chimpanzees are Indifferent to the Welfare of Unrelated Group Members.' Joan B. Silk et al., *Nature*, Vol. 437 (27 October 2005), pp. 1357–9.

'Spontaneous Altruism by Chimpanzees and Young Children.' Felix Warneken et al., *PLoS Biology*, Vol. 5 (July 2007).

'Other-Regarding Preferences in a Non-Human Primate: Common Marmosets Provision Food Altruistically.' Judith M Burkart et al., *Proceedings of the National Academy of Sciences*, published online 5 December 2007, 10.1073/pnas. 0710310104.

'Causal Knowledge and Imitation/Emulation Switching in Chimpanzees and Children.' V. Horner and A. Whiten, *Animal Cognition*, Vol. 8 (July 2005), pp. 164–81.

'The Hidden Structure of Overimitation.' Derek E. Lyons et al., *Proceedings of the National Academy of Sciences*, Vol. 104 (11 December 2007), pp. 19751–6.

'The Evolutionary and Developmental Foundations of Mathematics.' Michael J. Beran, *PLoS Biology*, Vol. 6 (February 2008).

'Human Evolution: Why We're Different: Probing the Gap Between Apes and Humans.' Michael Balter, *Science*, Vol. 319 (25 January 2008), pp. 405–6.

Aping Language. Joel Wallman (Cambridge University Press, 1992).

'The Uniqueness of Human Recursive Thinking.' Michael C. Corballis, *American Scientist*, Vol. 95 (2007), pp. 240–8.

'Scientist Postulates 4 Aspects of 'Humaniqueness' Differentiating Human and Animal Cognition.' *Eurekalert*, 17 February 2008.

'The Roots of Morality.' Greg Miller, *Science*, Vol. 320 (9 May 2008), pp. 734–7.

'Darwin's Mistake: Explaining the Discontinuity between Human and Nonhuman Minds.' Derek C. Penn, Keith J. Holyoak, and Daniel J. Povinelli, *Behavioural and Brain Sciences*, Vol. 31 (2008), pp. 109–30.

Kanzi: The Ape at the Brink of the Human Mind. Sue Savage-Rumbaugh and Roger Lewin (New York: John Wiley and Sons, 1994).

Rogue Primate: An Exploration of Human Domestication. John A. Livingston (Toronto: Key Porter Books, 1994).

Touched with Fire: Manic-Depressive Illness and the Artistic Temperament. Kay Redfield Jamison (New York: Free Press Paperbacks, 1994).

INDEX